S0-ARJ-900

SRA

# Connecting Math Concepts

LEVEL
D

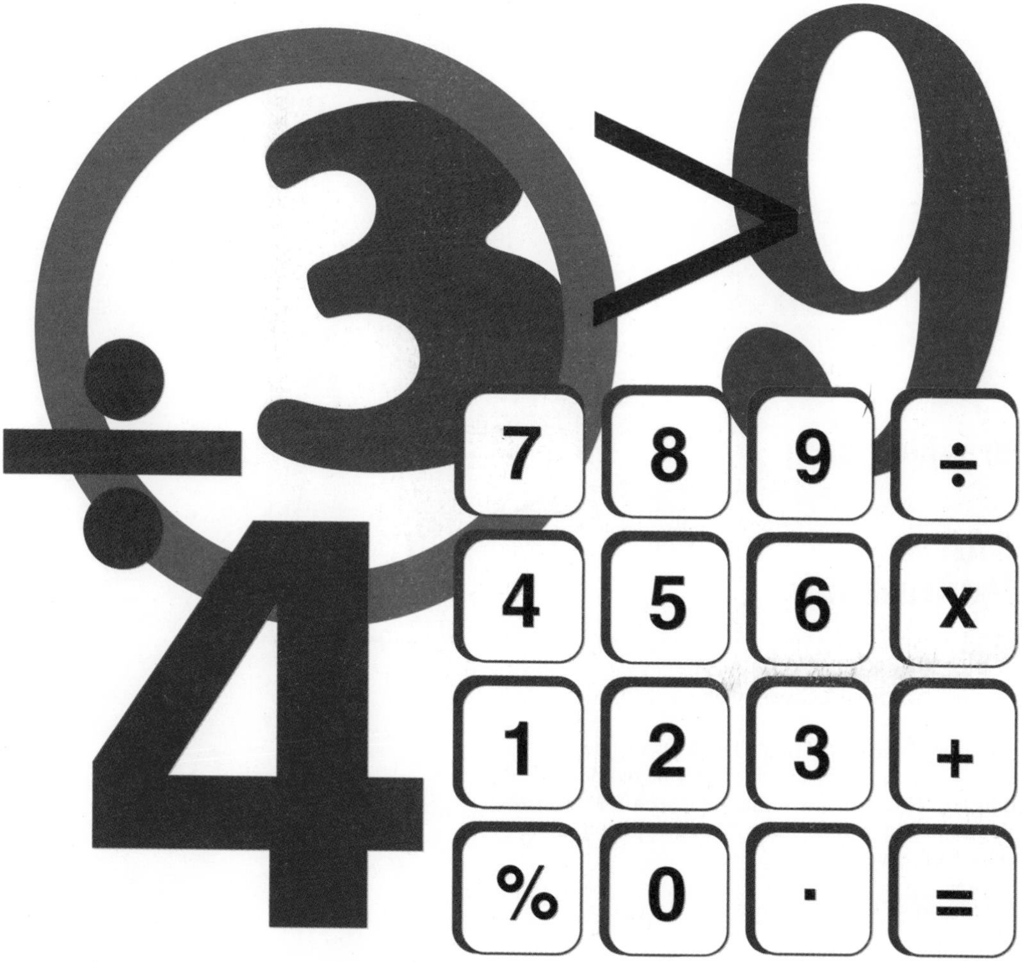


SRA
*Columbus, Ohio*

The McGraw·Hill Companies

www.sra4kids.com

SRA
McGraw-Hill

Copyright © 2003 by SRA/McGraw-Hill.

All rights reserved. Except as permitted under the United States Copyright Act, no part of this publication may be reproduced or distributed in any form or by any means, or stored in a database or retrieval system, without the prior written permission of the publisher, unless otherwise indicated.

Send all inquiries to:
SRA/McGraw-Hill
4400 Easton Commons
Columbus, OH 43219

Printed in the United States of America.

ISBN 0-02-684692-6

11 12 13 14 DOW 13 12 11 10

The McGraw·Hill Companies

# Connecting Math Concepts

## Level D

***Connecting Math Concepts*** will teach you a great deal about doing mathematics. When something new is introduced, your teacher will help you work the problems. Very soon, you'll be doing the problems on your own, without help.

Remember, everything that is introduced is something you will learn. It is something that you'll need to work even more difficult problems that will be introduced later.

Learning is easier when you follow your teacher's directions. Sometimes your teacher will direct you to work part of a problem, sometimes a whole problem, and sometimes a group of problems.

Listen very carefully to the directions and follow them. Work quickly and accurately. Most important of all, work hard. You'll be rewarded with math skills that will surprise you.

# Lesson 1

## Part 1

- You'll work problems that are presented in the textbook. Some are column problems like these:

a. 324
   – 110

b. 624
   – 104

- When you work column problems on lined paper, follow these rules:

  **1) Use the back of your paper.**
  **2) Turn your paper sideways.**
  **3) Write the letter of the problem.**
  **4) Copy the problem carefully. Copy it so that the digits are separated by the lines of your paper.**
  **5) Leave room between your problems. Don't crowd your problems together.**

- When you follow these rules, you'll be able to see what you've written. It's easier to keep your numbers lined up. You can find mistakes more easily.
- Here's problem A worked: Here's problem B worked:

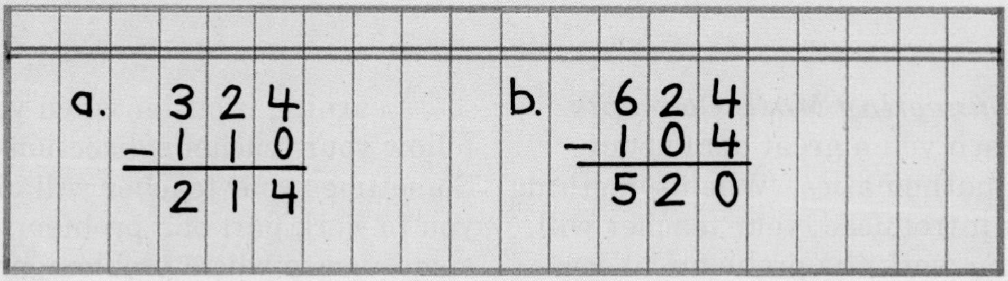


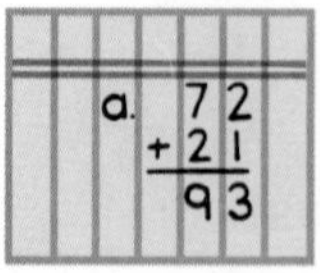


**Follow the rules and work these problems.**

a. 128
   + 224

b. 319
   + 306

c. 712
   + 152

## Part 2

a.

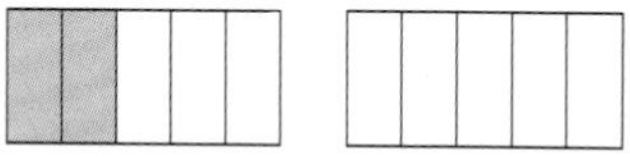

b.

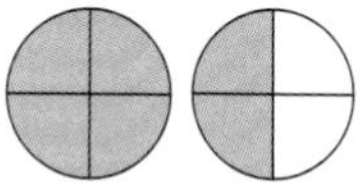

c.

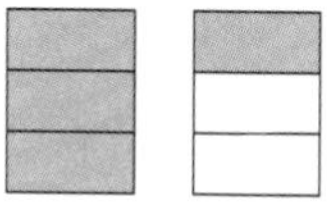

d.

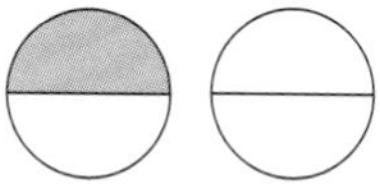

e.

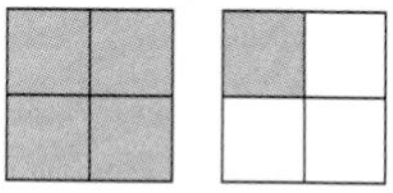

f.

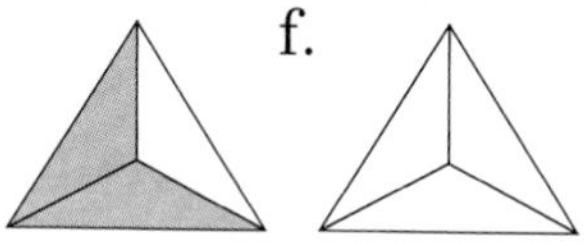

I'll have $\frac{4}{3}$ pie. And give $\frac{1}{3}$ to my brother.

## Part 3

- For some subtraction problems, you have to borrow.
- You have to borrow when you can't work a subtraction problem in a column.
- You start with the ones column and read the problem. If it begins with a smaller number, you can't work it.
- And if you can't work it, you borrow.

- Here's a problem:
- The problem in the ones column is 0 – 4.

$$\begin{array}{r} 5\,7\,0 \\ -\,1\,0\,4 \\ \hline \end{array}$$

- You can't work it, so you borrow 1 ten from the tens digit of 570.

$$\begin{array}{r} 5\,\overset{6}{\cancel{7}}{}^{1}0 \\ -\,1\,0\,4 \\ \hline \end{array}$$

- Now you have a new problem in the ones column: 10 – 4. You also have a new problem in the tens column: 6 – 0. You can work both those problems and the problem in the hundreds column.

$$\begin{array}{r} 5\,\overset{6}{\cancel{7}}{}^{1}0 \\ -\,1\,0\,4 \\ \hline 4\,6\,6 \end{array}$$

- Sometimes you have to borrow to work the problem in the tens column. You borrow from the **hundreds.**

- Here's a problem with the ones column already worked. The problem in the tens column is 7 – 8.

$$\begin{array}{r} 6\,7\,0 \\ -\,4\,8\,0 \\ \hline 0 \end{array}$$

- You can't work it, so you borrow from the hundreds digit of 670.

$$\begin{array}{r} \overset{5}{\cancel{6}}{}^{1}7\,0 \\ -\,4\,8\,0 \\ \hline 0 \end{array}$$

- The new problem in the tens column is 17 – 8. The new problem in the hundreds column is 5 – 4. You can work both those problems.

$$\begin{array}{r} \overset{5}{\cancel{6}}{}^{1}7\,0 \\ -\,4\,8\,0 \\ \hline 1\,9\,0 \end{array}$$

## Part 4

a. 57
$-$ 39

b. 608
$-$ 328

c. 732
$-$ 227

d. 436
$-$ 182

## Part 5

- Some thousands numerals have 4 digits. To read these thousands numerals, you say **thousand** after the first digit. Then you read the rest of the numeral.
- Here's a numeral: **7 3 5 6**
- The first digit is 7. So you say **7 thousand.** Then you read the rest of the numeral: **3 hundred 56.**

**Read the numerals.**

a. 4 3 8 1

b. 6 1 1 9

c. 1 6 0 4

d. 3 2 0 0

No, no. I ordered 70 books, not 70 cases of books.
100 BOOKS

## Part 6

- A lot of problems you'll work are based on number families. A number family is made up of three numbers that always go together to make addition facts and subtraction facts.
- The family is made up of two small numbers and a big number:

  **small number** **small number** ⟶ **big number**

  The big number is at the end of the arrow.
- Here's a family with the small numbers shown: 4 6 ⟶ ■
- To find the big number, you add the small numbers: 4 + 6 = 10. The big number is 10. 4 + 6 = 10
- Here's a family with a small number missing: ■ 6 ⟶ 10
- To find a missing small number, you subtract. You start with the big number and subtract the small number that is shown: 10 – 6. The answer is 4. So the missing small number is 4. 10 – 6 = 4
- Here's a family with the other small number missing: 4 ■ ⟶ 10
- You subtract to find that number. The subtraction for this family is 10 – 4. The missing small number is 6. 10 – 4 = 6
- You can use number families to figure out answers to very difficult word problems.

a. 6 8 ⟶ ■

b. ■ 6 ⟶ 17

c. 150 ■ ⟶ 321

d. 67 135 ⟶ ■

e. ■ 76 ⟶ 89

f. 23 71 ⟶ ■

# Lesson 2

## Part 1

a. 9804

b. 2760

c. 5672

d. 4006

e. 6007

f. 5003

g. 6040

h. 6000

## Part 2

a. 47 ■ → 98

b. 29 56 → ■

c. ■ 104 → 206

d. 53 ■ → 79

e. 390 120 → ■

## Part 3

a. b. c. d. e.

That's the silliest number family I ever saw.
3
7
10

## Part 4

- For some addition problems, you have to carry. You work in the ones column first.

- For this problem, you add 7 + 5.

$$\begin{array}{r} 47 \\ \underline{+\,25} \end{array}$$

- The answer is 12. That's a 2-digit value. So you carry the first digit to the tens column. The second digit is the answer for the ones column.

$$\begin{array}{r} {\scriptstyle 1}\phantom{7} \\ 47 \\ \underline{+\,25} \\ 2 \end{array}$$

- Then you add the values in the tens column and write the answer.

$$\begin{array}{r} {\scriptstyle 1}\phantom{7} \\ 47 \\ \underline{+\,25} \\ 72 \end{array}$$

- Multiplication works the same way.
- Here's 47 x 5.
  The arrows show that you work two problems. You work the problem in the ones column first: 7 x 5.

$$\begin{array}{r} 47 \\ \searrow\downarrow \\ \underline{\times\;\;5} \end{array}$$

- The answer is 35. That's a 2-digit value, so you carry the first digit to the tens. The second digit is the answer for the ones column.

$$\begin{array}{r} {\scriptstyle 3}\phantom{7} \\ 47 \\ \searrow\downarrow \\ \underline{\times\;\;5} \\ 5 \end{array}$$

- Now you work the tens. **First, you multiply. Then you add.** The multiplication problem for the tens is 4 x 5. That's 20. Then you add 3.
- That's 23.

$$\begin{array}{r} {\scriptstyle 3}\phantom{7} \\ 47 \\ \searrow\downarrow \\ \underline{\times\;\;5} \\ 235 \end{array}$$

- Here's another problem:

$$\begin{array}{r} 15 \\ \times\ \ 9 \\ \hline \end{array}$$

- You carry the first digit of 45. The second digit is the answer in the ones column.

$$\begin{array}{r} {\scriptstyle 4}\ \\ 15 \\ \times\ \ 9 \\ \hline 5 \end{array}$$

- Now you multiply for the tens.
  Then you add the number you carried to 9.
  That's 9 + 4.

$$\begin{array}{r} {\scriptstyle 4}\ \\ 15 \\ \times\ \ 9 \\ \hline 135 \end{array}$$

## Independent Work

**Part 5** **Copy each problem and work it.**

a. $\begin{array}{r} 52 \\ \times\ 4 \\ \hline \end{array}$

b. $\begin{array}{r} 71 \\ \times\ 5 \\ \hline \end{array}$

c. $\begin{array}{r} 73 \\ \times\ 2 \\ \hline \end{array}$

# Lesson 3

## Part 1

a. $\begin{array}{r} 464 \\ -\ 382 \\ \hline \end{array}$

b. $\begin{array}{r} 73 \\ -\ 19 \\ \hline \end{array}$

c. $\begin{array}{r} 308 \\ -\ 165 \\ \hline \end{array}$

d. $\begin{array}{r} 628 \\ -\ 609 \\ \hline \end{array}$

## Part 2

a. ■ 116 → 237

b. 209 174 → ■

c. 301 ■ → 402

d. ■ 200 → 306

e. 186 248 → ■

## Part 3

a.

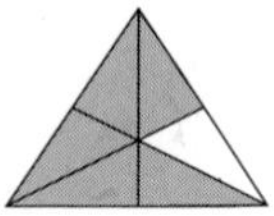

b.

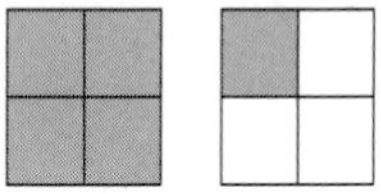

c.

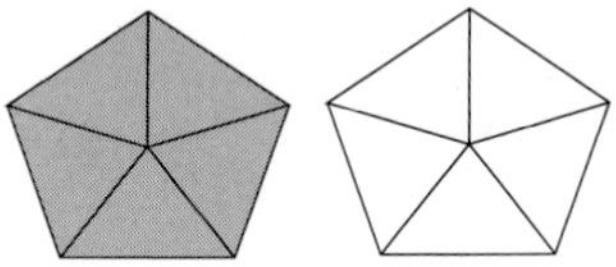

d.

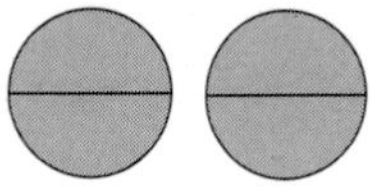

e.

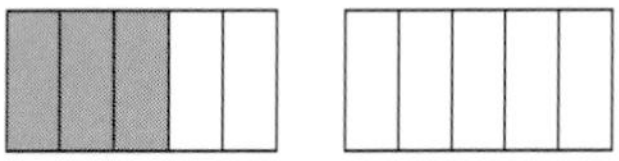

f.

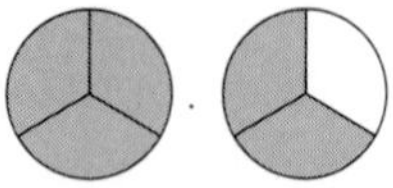

But you told me you'd be here before 9 o'clock.
12
1
2
3
4
5
6
7
8
9
10
11

## Part 4

- The **hour** hand on a clock is the short hand. The **minute** hand is the long hand.
- To figure out the time on a clock, you start with the hour hand. The **hour hand points to the number of hours.**
- The hour for this hand is 3:

- The hour for this hand is 8:

- The **minute hand does not point to the number of minutes.** You figure the minutes by starting at the top of the clock and counting by 5 for each number.
- Here's a clock with the minute numbers shown: The number for the minute hand is 20.

- To write the time, you first write the hour: **11**
- Then you make a colon. That's two dots: 11**:**
- Then you write the minutes: 11:**20**
- You can see the time shown for the clocks below.

8:05

1:15

4:55

## Part 5

a. b. c. d.

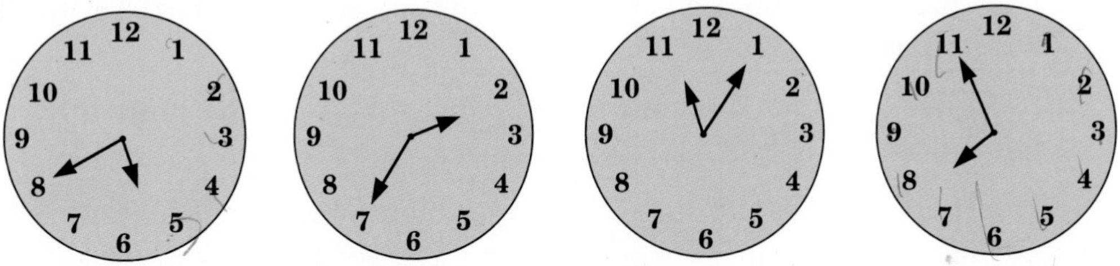


# Independent Work

## Part 6 Copy each problem and work it.

| a. | b. | c. | d. | e. | f. |
|---|---|---|---|---|---|
| 31 | 5 | 683 | 70 | 76 | 12 |
| 17 | 26 | – 161 | + 33 | – 33 | 120 |
| + 99 | + 8 | | | | + 279 |

## Part 7

| a. | b. | c. |
|---|---|---|
| 52 | 26 | 61 |
| x 4 | x 1 | x 5 |

## Part 8 Write the answer for each problem.

a. 2 x 7 = ■
b. 9 x 5 = ■
c. 9 x 2 = ■
d. 4 x 2 = ■
e. 9 x 0 = ■
f. 5 x 1 = ■
g. 5 x 4 = ■
h. 7 x 5 = ■
i. 5 x 10 = ■
j. 2 x 9 = ■
k. 2 x 7 = ■
l. 1 x 8 = ■

## Part J

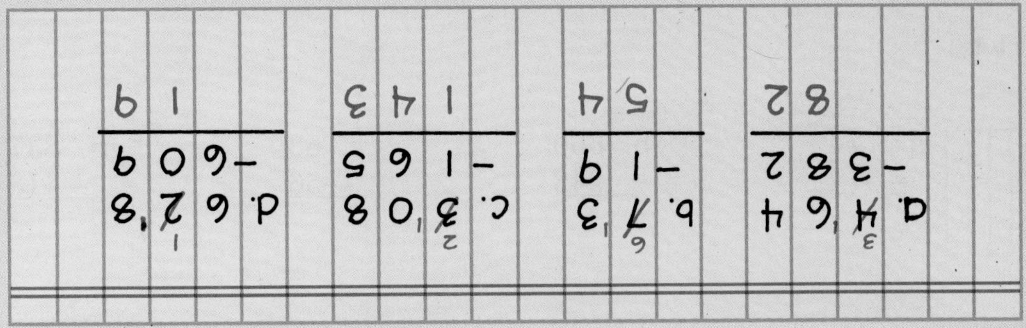

a. 464 − 382 = 82
b. 73 − 19 = 54
c. 308 − 165 = 143
d. 628 − 609 = 19

464 −382 142 73 −19 64
464 −382 82 73 −19 54
464 −382 182 73 −19 56
464 −382 82 73 −19 54
I think you got everything right.

# Lesson 4

## Part 1

a. $\begin{array}{r} 744 \\ -\ 372 \\ \hline \end{array}$

b. $\begin{array}{r} 650 \\ -\ 243 \\ \hline \end{array}$

c. $\begin{array}{r} 63 \\ -\ 56 \\ \hline \end{array}$

d. $\begin{array}{r} 837 \\ -\ 418 \\ \hline \end{array}$

## Part 2

a. b. c. d.

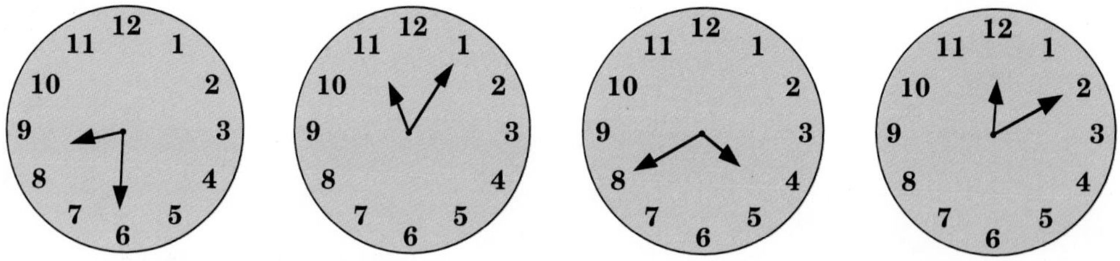


## Part 3

a. b. c. d. e.

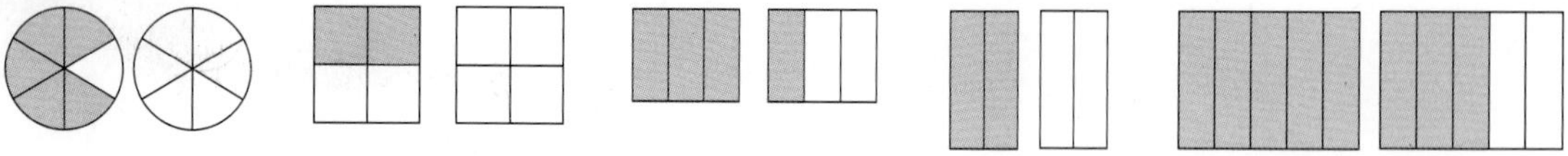

## Part 4

- You can use a calculator to check your work.
- Here's a problem a student worked. The student can check the problem by using a calculator.

$$\begin{array}{r} 474 \\ +\ 98 \\ \hline 582 \end{array}$$

- Here's what the student must enter in the calculator to get the answer to the problem.
- First, the student enters the digits for 474: [4] [7] [4]
- Next, the plus sign: [+]
- Next, the digits for 98: [9] [8]
- Last, the equal sign: [=]
- When the student presses [=], the calculator shows the answer to the problem. If the student forgets to press [=], the calculator doesn't do anything.

a. $\begin{array}{r} 474 \\ -\ 98 \\ \hline 376 \end{array}$

b. $\begin{array}{r} 319 \\ \times\ 8 \\ \hline 3040 \end{array}$

c. $\begin{array}{r} 903 \\ -\ 865 \\ \hline 38 \end{array}$

d. $\begin{array}{r} 784 \\ +\ 265 \\ \hline 1059 \end{array}$

## Part 5

| | |
|---|---|
| 5 | 10 |
| 15 | 20 |
| 25 | 30 |
| 35 | 40 |
| 45 | 50 |

| | |
|---|---|
| 1 | 2 |
| 3 | 4 |
| 5 | 6 |
| 7 | 8 |
| 9 | 10 |

## Part 6

a. 17 ■ → 56

b. 82 320 → ■

c. ■ 76 → 499

d. 27 72 → ■

# Independent Work

## Part 7

**Write the answer for each problem.**

a. 2 x 6 = ■
b. 2 x 8 = ■
c. 9 x 2 = ■
d. 5 x 2 = ■
e. 5 x 9 = ■
f. 2 x 7 = ■
g. 8 x 5 = ■
h. 8 x 2 = ■
i. 5 x 4 = ■
j. 7 x 1 = ■
k. 7 x 2 = ■
l. 5 x 5 = ■

## Part 8

**Copy the problems and work them.**

a. 56 + 5 + 17

b. 300 + 20 + 2

c. 6 + 50 + 30

d. 12 + 94 + 9

## Part 9

a. 27 x 2

b. 59 x 3

c. 82 x 5

d. 36 x 2

e. 63 x 2

f. 84 x 1

g. 31 x 9

## Part J

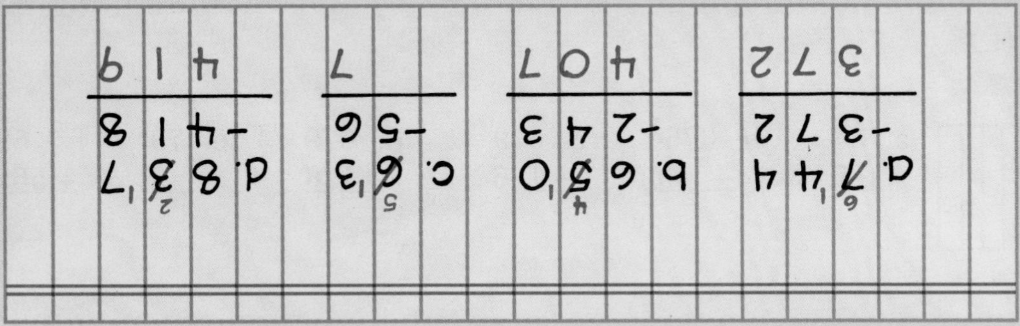

a. 744 −372 = 372
b. 650 −243 = 407
c. 63 −56 = 7
d. 837 −418 = 419

63 −56 = 17
63 −56 = 7
63 −56 = 7
63 −56 = 7
63 −56 = 7
63 minus 56 equals 17.
WRONG!

# Lesson 5

## Part 1

a. 34 x 5

b. 3764 − 45

c. 679 + 126

d. 679 − 126

e. 79 x 2

f. 52 + 569

## Part 2

a. b. c. d. e.

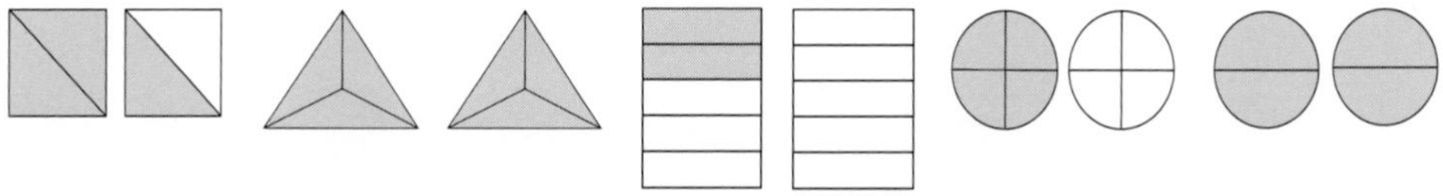

## Part 3

| | |
|---|---|
| 5 | |
| 15 | |
| 25 | |
| | 40 |
| | 50 |

| | |
|---|---|
| 1 | 2 |
| 3 | 4 |
| 5 | 6 |
| 7 | 8 |
| 9 | 10 |

## Part 4

a. 270 230 → ■

b. ■ 310 → 782

c. ■ 129 → 730

d. 29 ■ → 60

e. 58 76 → ■

f. 561 339 → ■

## Part 5

a.
```
   1 4
   2 1 7
 x     6
 1 3 1 2
```

b.
```
   1
   4 5 6
 + 8 2 5
 1 2 8 1
```

c.
```
   1 1
   5 0 7
 +   9 4
   6 0 1
```

d.
```
   6
   7̸ ¹4 2
 − 6 8 0
     6 2
```

## Part 6

| | |
|---|---|
| | 9 |
| 1 | 8 |
| 2 | 7 |
| 3 | 6 |
| 4 | 5 |
| 5 | 4 |
| 6 | 3 |
| 7 | 2 |
| 8 | 1 |
| 9 | 0 |

## Independent Work

## Part 7

**Copy each problem and work it.**

a.
```
    37
   406
 + 310
```

b.
```
    12
   721
 + 409
```

c.
```
   612
 + 105
```

d.
```
    23
   467
 +  19
```

## Part 8

**Write the answer for each problem.**

| a. | b. | c. | d. | e. | f. | g. |
|---|---|---|---|---|---|---|
| 1 | 8 | 8 | 6 | 14 | 9 | 5 |
| x 9 | x 2 | – 2 | x 0 | – 5 | x 2 | x 3 |

| h. | i. | j. | k. | l. | m. |
|---|---|---|---|---|---|
| 5 | 5 | 10 | 9 | 6 | 5 |
| x 1 | – 0 | x 4 | – 7 | x 2 | x 3 |

## Part 9

- The hour on a clock is the number that the short hand points to.
- The minutes on a clock are 5 times the number the long hand points to.
- To write the time for clocks, you first write the hour. Then you make a colon. Then you write the minutes.

**Write the time shown on each clock.**

a.

b.

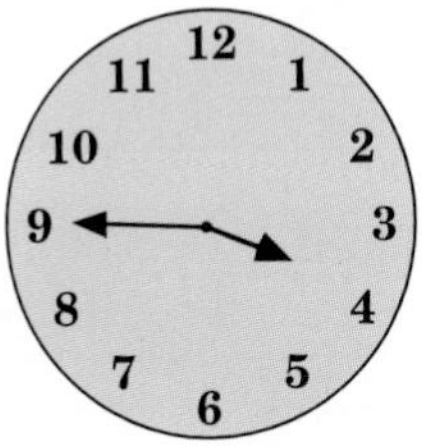


c.

d.

# Lesson 6

## Part 1

- Some addition problems have more than two numbers.
- Here's a problem that's already worked:

$$\begin{array}{r} {}^{1\,1}\phantom{0} \\ 642 \\ 87 \\ +205 \\ \hline 934 \end{array}$$

- Here's how you check the answer with your calculator.

| | |
|---|---|
| First enter the digits for 642: | **6** **4** **2** |
| Then enter plus: | **+** |
| Then enter the digits for 87: | **8** **7** |
| Then enter plus: | **+** |
| Then enter the digits for 205: | **2** **0** **5** |
| Then enter equals: | **=** |

- The answer shown above is correct: 642 + 87 + 205 = 934.

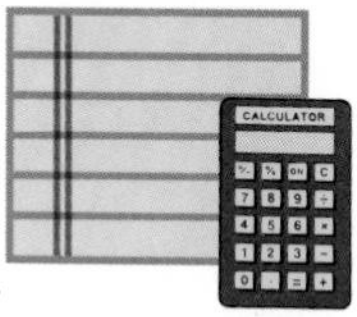

a. $\begin{array}{r} 16 \\ 372 \\ +\ 907 \\ \hline 1286 \end{array}$

b. $\begin{array}{r} 12 \\ 75 \\ +\ 190 \\ \hline 277 \end{array}$

c. $\begin{array}{r} 201 \\ 87 \\ +\ 759 \\ \hline 947 \end{array}$

**Part 2**

- You'll be doing work with number families that have an arrow pointing down. They work just like families that have an arrow going to the side.
- Here's a number family with a missing big number: 9 10 → ■
  The small numbers are 9 and 10.
- Here's the same family with the arrow pointing down: The missing big number is at the end of the arrow. The small numbers are 9 and 10. They are above the missing big number.

  9
  10
  ↓
  ■

- You add 9 and 10 to find the missing big number.

  9
  10
  ↓
  19

- Here's another number family:

  ■
  17
  ↓
  75

- Here's another number family:

  92
  ■
  ↓
  130

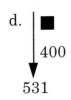

**Part 3** **Figure out the missing number for each family.**

a. 174, 315 ↓ ■

b. 127, ■ ↓ 197

c. 88, 65 ↓ ■

d. ■, 400 ↓ 531

## Part 4

| | |
|---|---|
| | 10 |
| | 20 |
| | |
| 35 | |
| 45 | |

| | |
|---|---|
| 1 | 2 |
| 3 | 4 |
| 5 | 6 |
| 7 | 8 |
| 9 | 10 |

## Part 5

a.

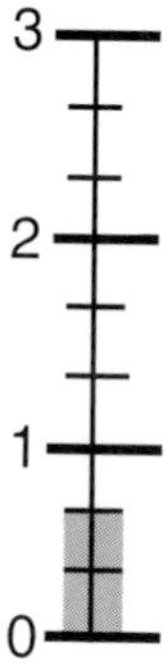

3
2
1
0

b.

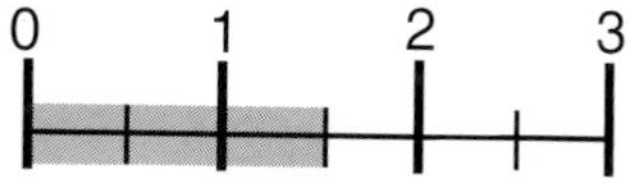

0
1
2
3

c.

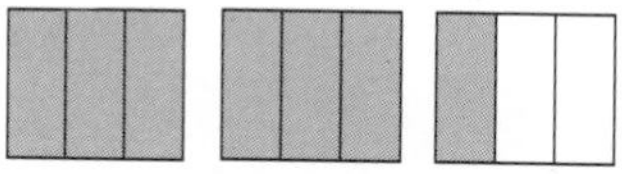

d.

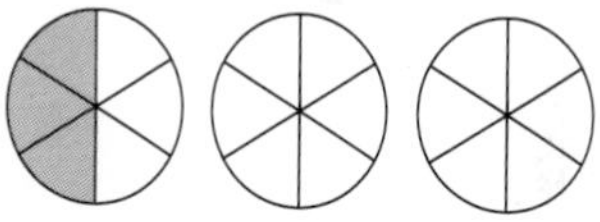

e.

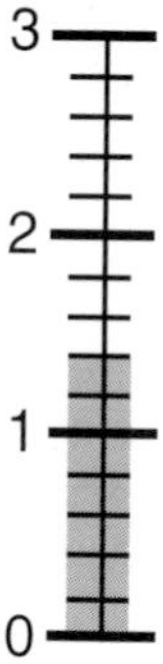

3
2
1
0

I thought they were supposed to eat $\frac{3}{7}$ of the pizza.

## Part 6

- Here's a problem that a student worked:

$$\begin{array}{r} {}^{2}\phantom{0} \\ 64 \\ \times\ \ 5 \\ \hline 400 \end{array}$$

- The answer is wrong.
- The student worked the problem in the ones column correctly.
- The student carried correctly.
- Then the student made a terrible mistake. The student didn't multiply 6 x 5. The student added 2 + 6. Then the student multiplied 8 x 5.
- Remember, first you multiply. Then you add the number you carried.
- Here's the problem worked correctly:

$$\begin{array}{r} {}^{2}\phantom{0} \\ 64 \\ \times\ \ 5 \\ \hline 320 \end{array}$$

## Independent Work

## Part 7 Figure out the missing number for each family.

a. ■ 42 → 706  b. 740 29 → ■  c. 650 ■ → 893

## Part 8 Write the answer for each problem.

a. $\begin{array}{r} 6 \\ \underline{\times 5} \end{array}$  b. $\begin{array}{r} 9 \\ \underline{\times 5} \end{array}$  c. $\begin{array}{r} 8 \\ \underline{\times 5} \end{array}$  d. $\begin{array}{r} 4 \\ \underline{\times 2} \end{array}$  e. $\begin{array}{r} 9 \\ \underline{\times 2} \end{array}$  f. $\begin{array}{r} 6 \\ \underline{\times 2} \end{array}$  g. $\begin{array}{r} 5 \\ \underline{\times 9} \end{array}$

## Part 9

**Write the time shown on each clock.**

a.

b.

c.

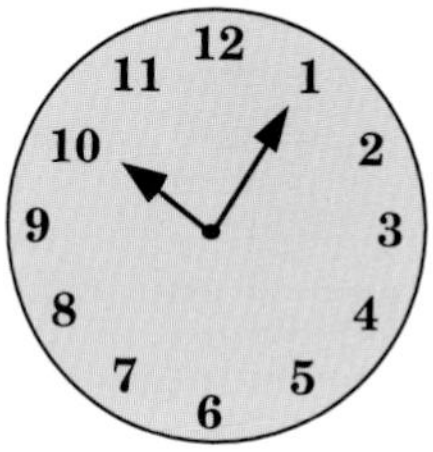

## Part 10

**Copy each problem and work it.**

a. $\begin{array}{r} 608 \\ -\ 210 \\ \hline \end{array}$

b. $\begin{array}{r} 476 \\ -\ 186 \\ \hline \end{array}$

c. $\begin{array}{r} 235 \\ -\ 119 \\ \hline \end{array}$

d. $\begin{array}{r} 963 \\ -\ 872 \\ \hline \end{array}$

## Part J

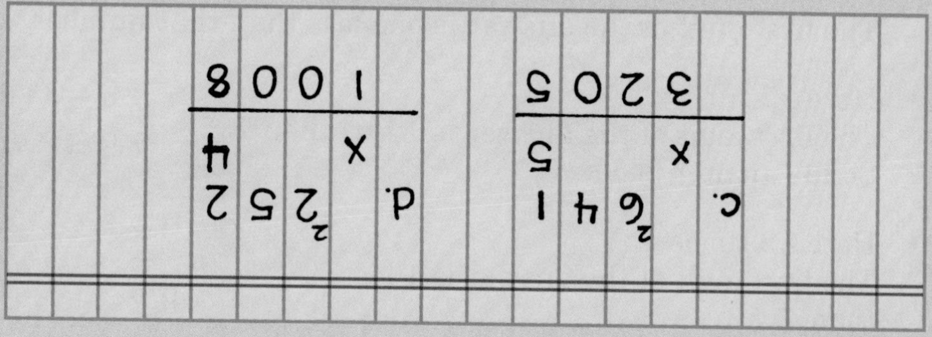

# Lesson 7

## Part 1

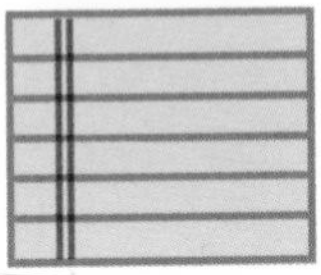

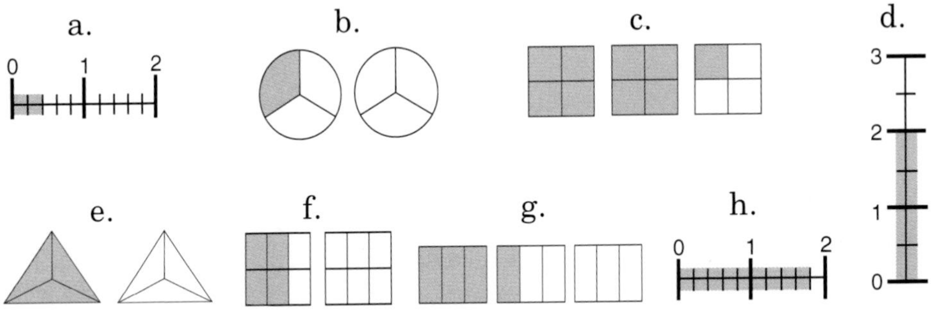


## Part 2

- When one of the numbers you multiply is 9, you can figure out the answer by thinking of the number map for counting by 9.
- Here's how you do it: You look at the number that is **not 9.** The first digit of the answer is one less than that number.
- Here's 7 times 9:
  The first digit of the answer is 1 less than 7.
  So the answer is 63:

  7 x 9 =
  7 x 9 = 6 3

- Here's 9 times 4:
  The first digit of the answer is 1 less than 4.
  So the answer is 36:

  9 x 4 =
  9 x 4 = 3 6

| | |
|---|---|
| | 9 |
| 1 | 8 |
| 2 | 7 |
| 3 | 6 |
| 4 | 5 |
| 5 | 4 |
| 6 | 3 |
| 7 | 2 |
| 8 | 1 |
| 9 | 0 |

- Remember, look at the number that is not 9. The first digit of the answer is one less than that number.

## Part 3

| | |
|---|---|
| 1 | 2 |
| 3 | 4 |
| 5 | 6 |
| 7 | 8 |
| 9 | 10 |

## Part 4

- Some thousands numerals have more than four digits. When a thousands numeral has more than four digits, there is a comma after the thousands.

- Here are thousands numerals with more than four digits:

  **35,000**
  **123,200**
  **86,050**

- Reading these numerals is not difficult. You read to the comma, then say **thousand,** then read the rest of the numeral.

  **35 thousand**
  **123 thousand 2 hundred**
  **86 thousand 50**

- Remember, read to the comma, say **thousand,** then read the rest of the numeral.

**Read the numerals.**

a. 62,800
b. 119,020
c. 300,640
d. 57,072

## Part 5

a. 10, ■ → 55

b. 72, 86 → ■

c. ■, 516 → 838

d. ■, 65 → 92

e. 345, 543 → ■

## Independent Work

## Part 6

**Write the time shown on each clock.**

a.

b.

c.

d.

## Part 7

**Figure out the missing number for each family.**

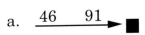

a. 46 91 → ■

b. 12 ■ → 89

c. ■ 39 → 56

## Part 8

**Copy each problem and work it.**

a. $\begin{array}{r} 57 \\ \underline{\times\ 3} \end{array}$

b. $\begin{array}{r} 25 \\ \underline{\times\ 9} \end{array}$

c. $\begin{array}{r} 86 \\ \underline{\times\ 2} \end{array}$

d. $\begin{array}{r} 123 \\ \underline{\times\ \ 2} \end{array}$

## Part 9

a. 375
 − 185

b. 84
 − 19

c. 873
 − 519

d. 542
 − 233

## Part 10

**Write the answer to each problem.**

a. 8 x 5 = ■

b. 2 x 8 = ■

c. 1 x 5 = ■

d. 6 x 2 = ■

e. 5 x 5 = ■

f. 2 x 7 = ■

g. 2 x 8 = ■

h. 9 x 1 = ■

# Lesson 8

## Part 1

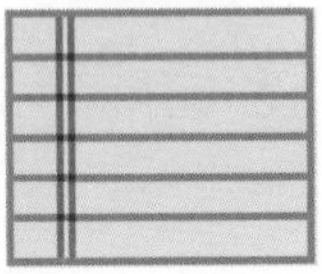

a.

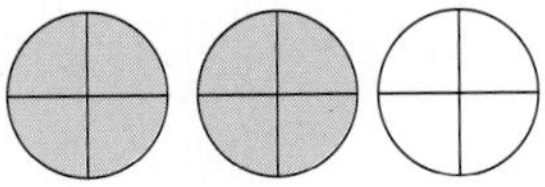

b.

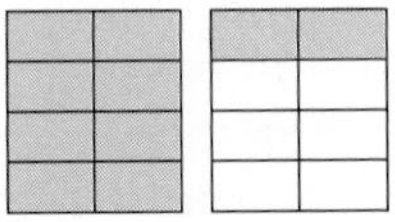

c.

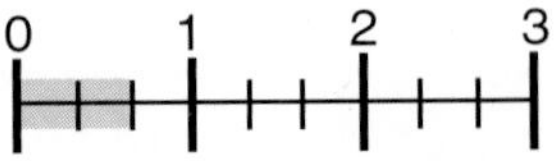


d.

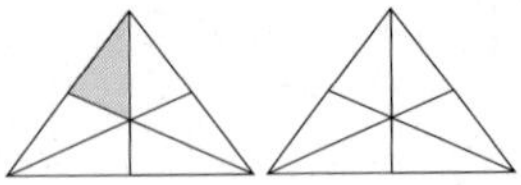

e.

f.

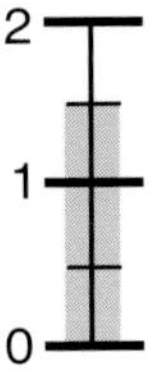


## Part 2

| | |
|---|---|
| 1 | 2 |
| 3 | 4 |
| 5 | 6 |
| 7 | 8 |
| 9 | 10 |

## Part 3

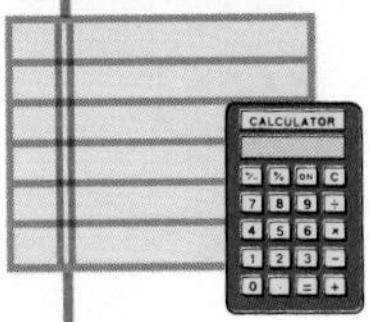


a. $\begin{array}{r} 860 \\ 709 \\ \underline{+\ 573} \\ 2042 \end{array}$

b. $\begin{array}{r} 906 \\ \underline{-\ \ 87} \\ 819 \end{array}$

c. $\begin{array}{r} 684 \\ \underline{+\ 537} \\ 1221 \end{array}$

d. $\begin{array}{r} 416 \\ \underline{\times\ \ 17} \\ 433 \end{array}$

## Part 4

- You can write 12 dollars and 7 cents using symbols. You follow these rules:
- Start with the dollar sign: **$**
- Write the number of dollars: $**12**
- Write a decimal point for the word **and:** $12**.**
- Then write the **2-digit** number for the cents: $12.**07**
- Remember: dollar sign, number of dollars, decimal point, and two digits for the cents.

a. 9 dollars and 6 cents

b. 36 dollars and 9 cents

c. 4 dollars and 17 cents

d. 20 dollars and 70 cents

e. 3 dollars and 99 cents

## Part 5

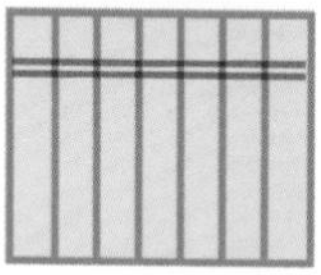

a. $\begin{array}{r} 725 \\ -\ 198 \\ \hline \end{array}$

b. $\begin{array}{r} 680 \\ -\ 186 \\ \hline \end{array}$

c. $\begin{array}{r} 837 \\ -\ 388 \\ \hline \end{array}$

## Independent Work

**Part 6** **These are multiplication, subtraction and addition problems. Copy each problem and work it. Make sure you copy the correct sign for each problem.**

a. 592
x 4

b. 592
− 477

c. 25
643
+ 58

d. 318
x 5

e. 318
− 167

f. 63
x 5

g. 285
+ 9

**Part 7** **Write the answer to each problem.**

a. 1 x 4 = ■
b. 9 x 2 = ■
c. 9 x 7 = ■
d. 9 x 5 = ■

e. 9 x 3 = ■
f. 9 x 9 = ■
g. 9 x 6 = ■
h. 5 x 4 = ■

i. 5 x 8 = ■
j. 5 x 2 = ■
k. 8 x 2 = ■

# Lesson 9

## Part 1

a. 326,009 b. 26,907 c. 64,500 d. 105,240

## Part 2

- Multiplication is just a faster way to add. You can use multiplication whenever the same number is added again and again.
- Here's a problem: How many squares are in the rectangle?

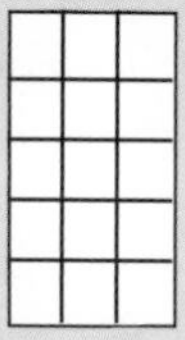

- You can use addition to solve the problem. Or you can use multiplication to solve the problem faster.

- Here's how you solve the problem by using addition: First you count the number of squares in the first column. There are 5 squares in the first column. So you write 5.

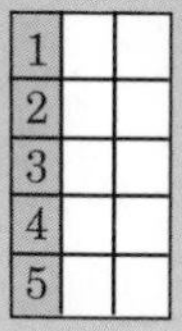


**5**

- Then you add the number for the second column. That's 5.

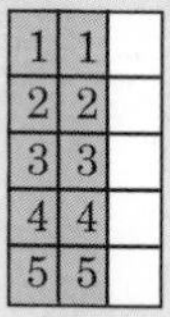


5 + **5**

- Then you do the same thing for the last column:

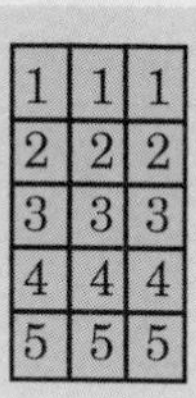


5 + 5 + **5** = **15**

- There are 15 squares in the rectangle.

- To solve the problem using multiplication, you start the same way. You write the number of squares in the first column:

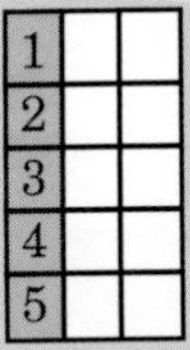


**5**

- Then you multiply by the number of columns. There are 3 columns. So you multiply by 3:

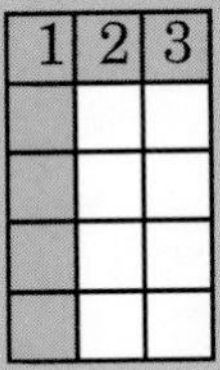


**5 x 3 = 15**

- You get the same answer: 15.
  There are 15 squares in the rectangle.

- Here's a different rectangle.
  There are 2 squares in each column. So you start with 2:

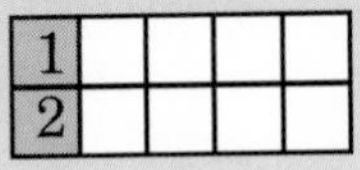


**2**

- To find the number of squares by adding, you add 2 for each of the other columns:

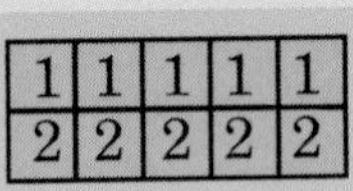


**2 + 2 + 2 + 2 + 2 = 10**

- The answer is 10.
- When you work the problem using multiplication, you start with the number in the first column:

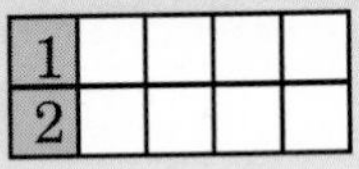


**2**

- Then you multiply by the number of columns:

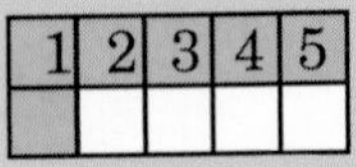


**2 x 5 = 10**

- There are 5 columns. You get the same answer: 10. There are 10 squares in the rectangle.

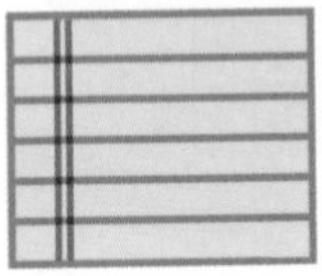

a. 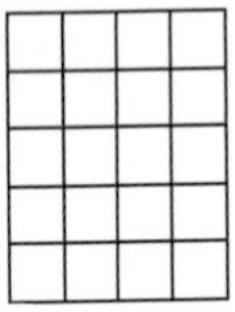

b. 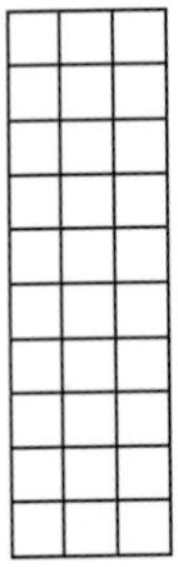

## Part 3

a. $\begin{array}{r} 656 \\ -\ 370 \\ \hline \end{array}$

b. $\begin{array}{r} 514 \\ -\ 206 \\ \hline \end{array}$

c. $\begin{array}{r} 24 \\ -\ \ 8 \\ \hline \end{array}$

d. $\begin{array}{r} 372 \\ -\ 294 \\ \hline \end{array}$

## Part 4

a.

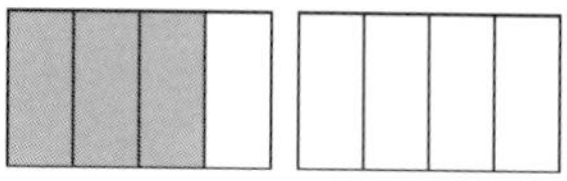

b.

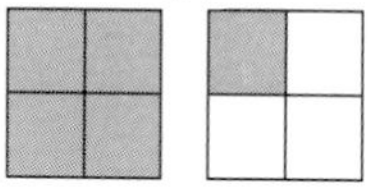

c.

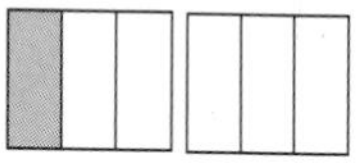

d.

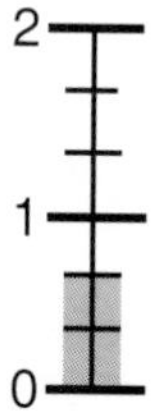


## Part 5

a. 14 dollars and 1 cent

b. 3 dollars and 72 cents

c. 12 dollars and 10 cents

d. 11 dollars and 98 cents

## Part 6

- Students sometimes make mistakes on multiplication problems that have 1 or zero. The students wouldn't have a problem if they thought about what the problem told them to do.

- Here's 5 x 1:
- That tells them to count by 5 one time. The answer is 5.

5 x 1

5 x 1 = 5

- Here's 5 x 0:
- That tells them to count by 5 zero times. That's no times at all. If you count no times at all, you'll end up at zero.

5 x 0

5 x 0 = 0

- Here's the rule: **Any number times zero equals zero.**

## Independent Work

### Part 7

**Copy each problem and work it.**

a. $\begin{array}{r} 56 \\ \underline{\times\ 9} \end{array}$

b. $\begin{array}{r} 408 \\ \underline{\times\ \ \ 5} \end{array}$

c. $\begin{array}{r} 701 \\ \underline{\times\ \ \ 9} \end{array}$

d. $\begin{array}{r} 270 \\ \underline{\times\ \ \ 2} \end{array}$

### Part 8

a. $\begin{array}{r} 236 \\ \underline{+\ 189} \end{array}$

b. $\begin{array}{r} 246 \\ \underline{+\ 806} \end{array}$

c. $\begin{array}{r} 178 \\ \underline{+\ 178} \end{array}$

d. $\begin{array}{r} 19 \\ 523 \\ \underline{+\ \ 41} \end{array}$

### Part 9

**Write the time shown on each clock.**

a.

b.

c.

d.

### Part J

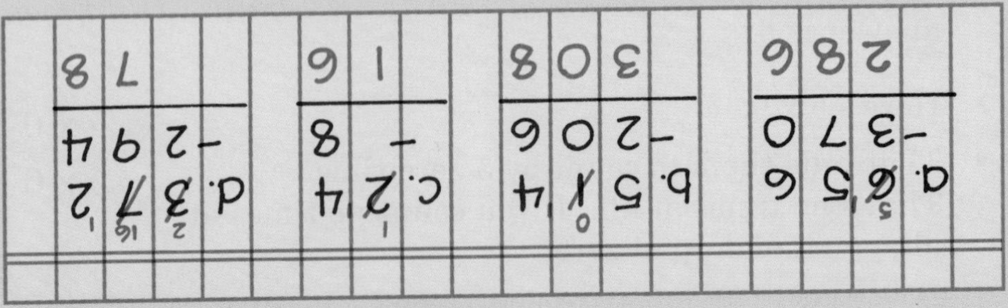

# Lesson 10

## Part 1

| | |
|---|---|
| 1 | 2 |
| 3 | 4 |
| 5 | 6 |
| 7 | 8 |
| 9 | 10 |

## Part 2

a. 3 ■ 2 = 6

b. 4 + ■ = 7

c. ■ – 1 = 8

d. 5 ■ 3 = 2

e. 8 ■ 4 = 12

f. ■ + 2 = 10

g. 3 x ■ = 15

h. 6 ■ 3 = 3

If you get 4 test items wrong, and 9 items right, how many test items are there?

# Lesson 11

## Part 1

- If a fraction is a whole number, the picture of the fraction shows units that are completely shaded and have no parts left over.
- **These pictures show whole numbers.**
- 3 **units** are shaded. So the fraction for this picture equals 3.

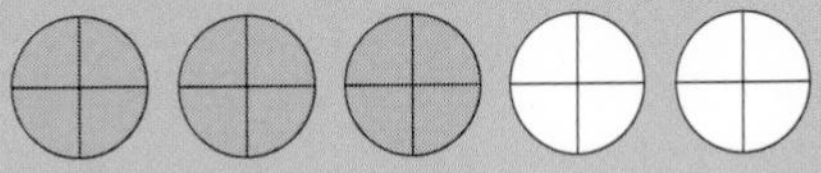

- 5 **units** are shaded. So the fraction for this picture equals 5.

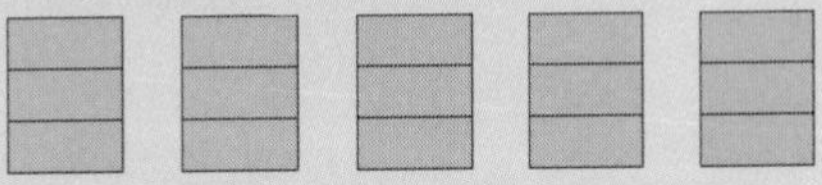

- 2 **units** are shaded. So the fraction for this picture equals 2.

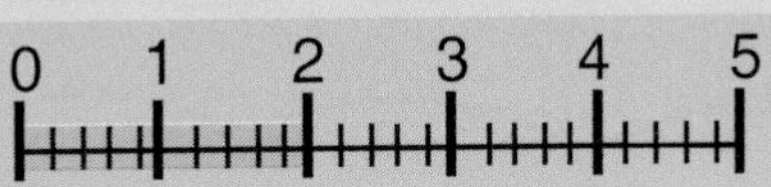


- **These pictures do not show whole numbers.**
- The fraction for this picture is more than 3 and less than 4.

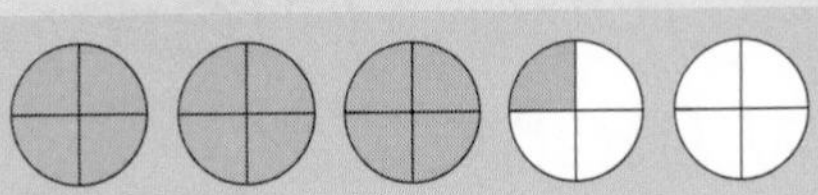

- The fraction for this picture is more than 5 and less than 6.

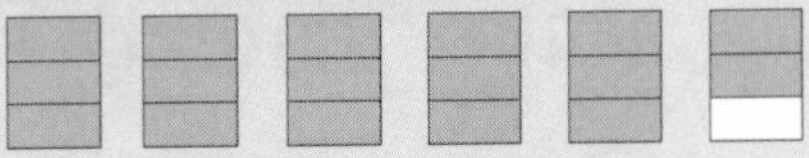

- The fraction for this picture is more than 1 and less than 2.

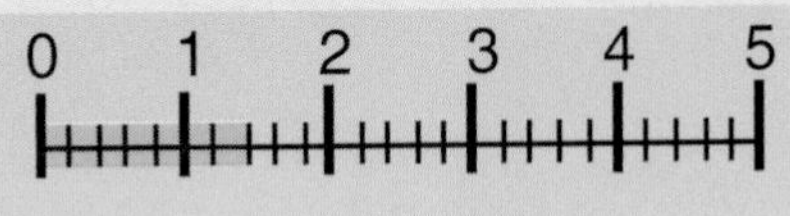


- Remember, if a fraction equals a whole number, **no** leftover parts are shaded.

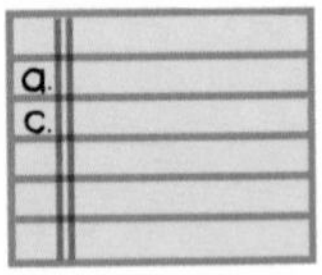


a. 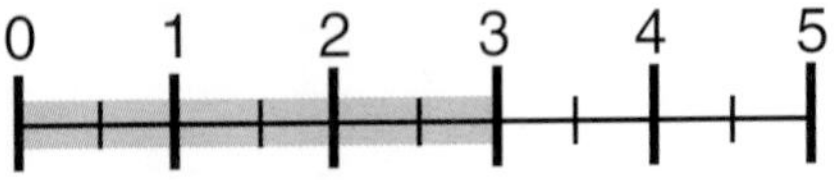


b. 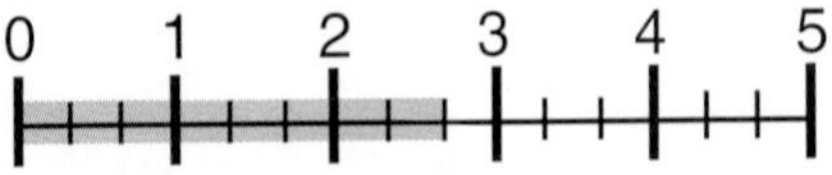


c. 

d. 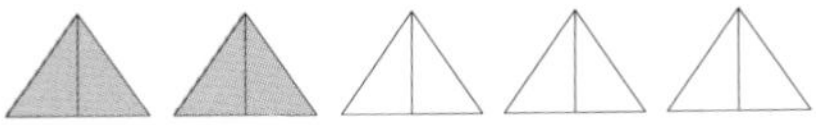

e. 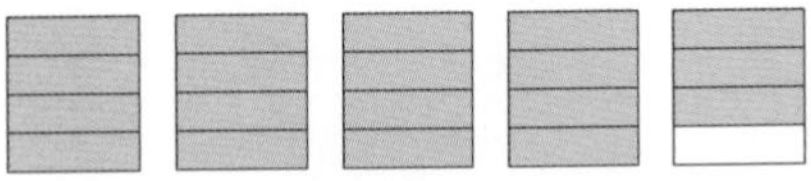

f. 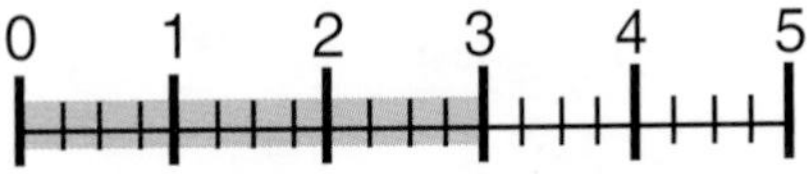

## Part 2

This is a number map for counting by 3. For numbers that are not in the top row, the second digit of each number is 1 less than the digit above it.

| | | |
|---|---|---|
| 3 | 6 | 9 |
| 12 | 15 | 18 |
| 21 | 24 | 27 |
| 30 | | |

## Part 3

a. 942 − 658

b. 36 − 19

c. 401 − 361

d. 556 − 358

## Independent Work

## Part 4

**Copy each problem and work it.**

a. 43 x 5

b. 74 x 9

c. 123 x 2

d. 567 x 2

e. 709 x 2

f. 57 + 381

g. 12 + 476 + 309

h. 88 + 999

## Part 5

**Copy each problem and write the answer.**

a. 9 x 2  b. 9 x 9  c. 9 x 6  d. 1 x 9  e. 5 x 9

f. 9 x 8  g. 9 x 7  h. 3 x 9  i. 4 x 9

## Part 6

**Write the column problem and the answer for each family. Make a box around each answer.**

a. 563 ■ → 999

b. ■, 120 → 287

c. 56, 296 → ■

d. ■ 40 → 90

## Part J

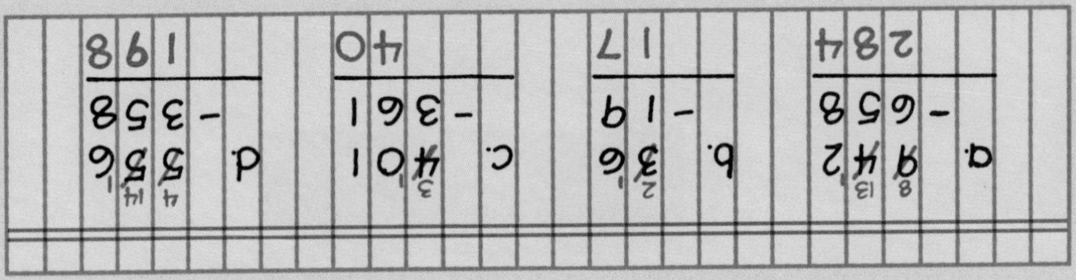

# Lesson 12

## Part 1

a. 132,400
b. 6,384
c. 12,056
d. 490,003
e. 9,103

## Part 2

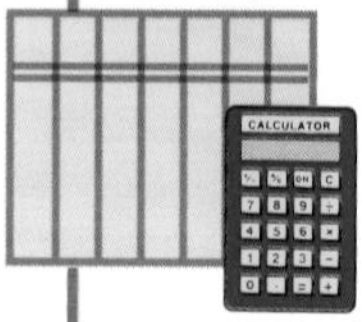


a. 674 − 293

b. 723 − 127

c. 76 − 72

d. 582 − 351

## Part 3

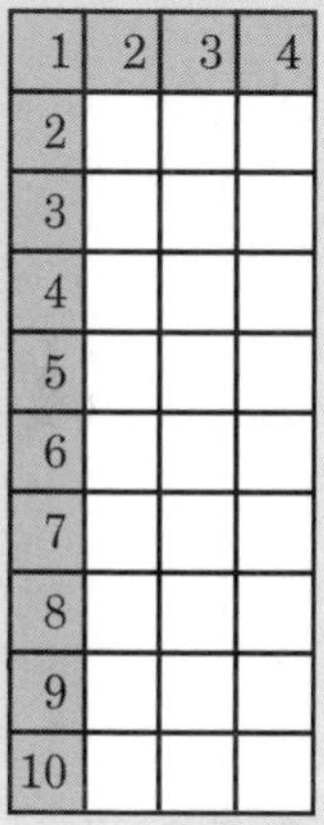


- You can figure out the number of squares in a rectangle by adding the column number again and again. Or you can do it the fast way by multiplying.
- There are 10 squares in each column. There are 4 columns.
- For the addition problem, you add 10 four times:

$$\begin{array}{r} 10 \\ 10 \\ 10 \\ +\ 10 \\ \hline 40 \end{array} \text{ squares}$$

- The answer is 40 squares.
- For multiplication, you start with the number of squares in a column and multiply by the number of columns. There are 4 columns. So you multiply 10 by 4:

$$\begin{array}{r} 10 \\ \times\ 4 \\ \hline 40 \end{array} \text{ squares}$$

- The answer is 40 squares.
- When you write answers to problems that deal with squares, write the name in the answer. You can write **squares** or you can write the abbreviation: **sq**

a. 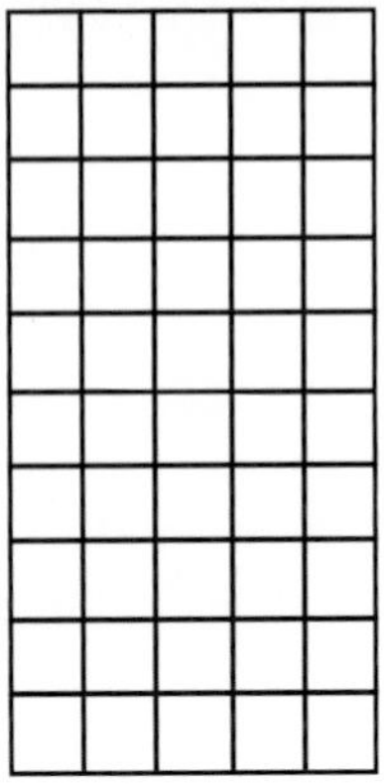

b. 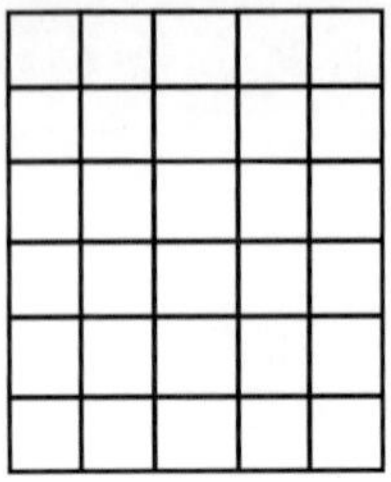

c. 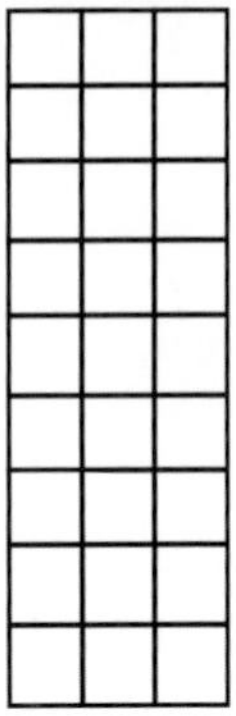

## Part 4

a. 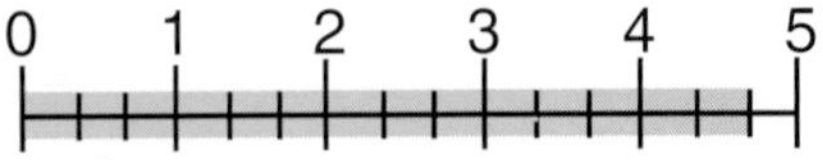


b. 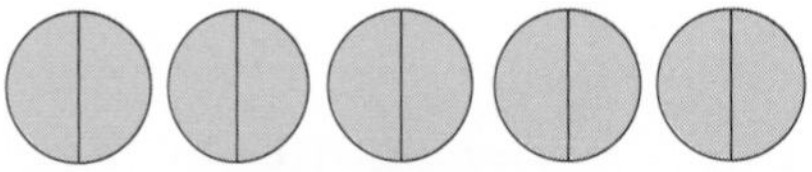

c. 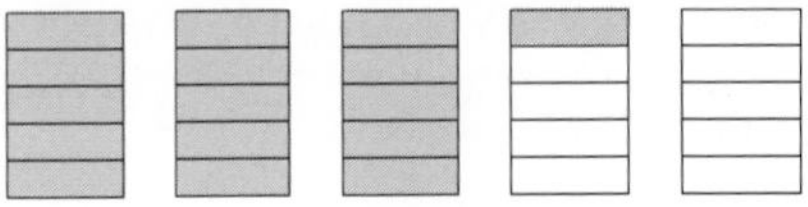

d. 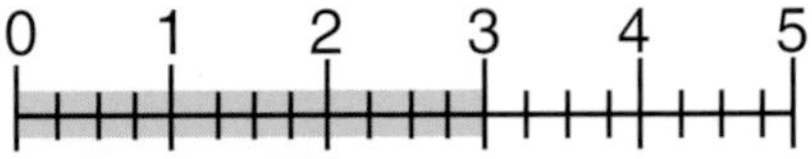


e. 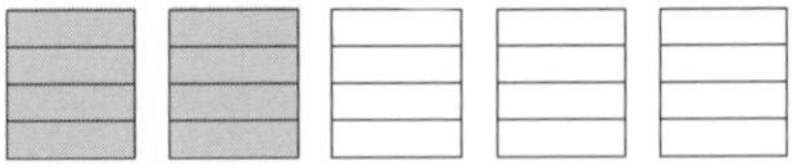

f. 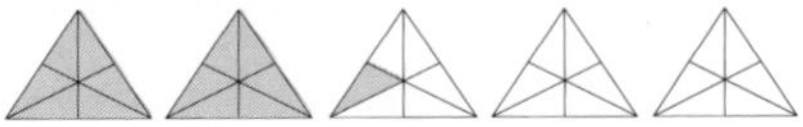

## Independent Work

## Part 5

**Copy each problem and work it.**

a. $\begin{array}{r} 346 \\ \underline{\times \quad 5} \end{array}$

b. $\begin{array}{r} 49 \\ \underline{+\ 625} \end{array}$

c. $\begin{array}{r} 12 \\ \underline{\times\ 9} \end{array}$

d. $\begin{array}{r} 315 \\ \underline{-\ 219} \end{array}$

e. $\begin{array}{r} 426 \\ \underline{\times \quad 9} \end{array}$

## Part 6 Write the fraction for each picture.

a.

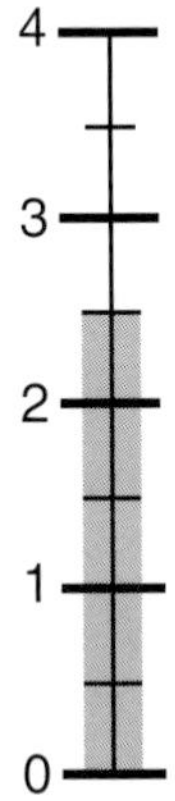


b.

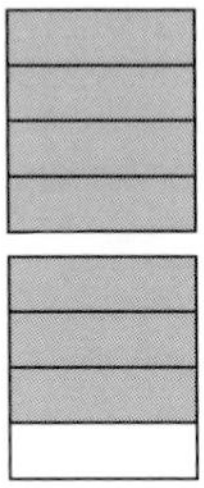

c.

d.

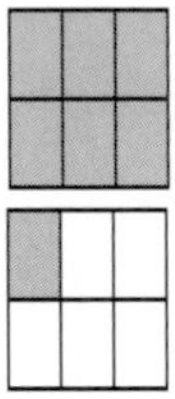

## Part 7 Write the time shown on each clock.

a.

b.

c.

Does one of these clocks say 9:30?
I don't know. I'll listen more closely.

# Lesson 13

## Part 1

a. 436 – 28 = ■

b. 563 – 96 = ■

c. 730 – 640 = ■

d. 814 – 75 = ■

## Part 2

- Some fractions for whole numbers have 1 part in each unit.
- Here's a picture: 

  There's 1 part in each unit. 4 parts are shaded. The fraction for the picture is 4 over 1.

  Here's the equation: $\frac{4}{1} = 4$
- Here's the picture of another fraction that has 1 part in each unit:

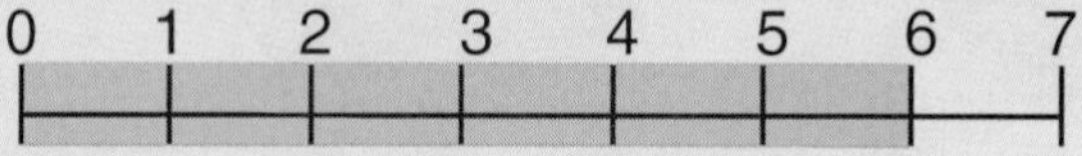


- 6 parts are shaded. Here's the equation: $\frac{6}{1} = 6$

a. 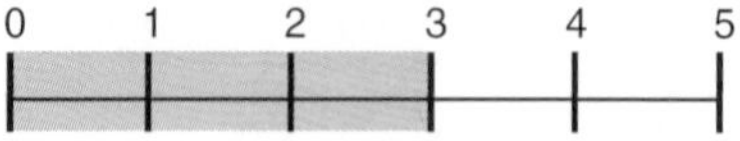


b. 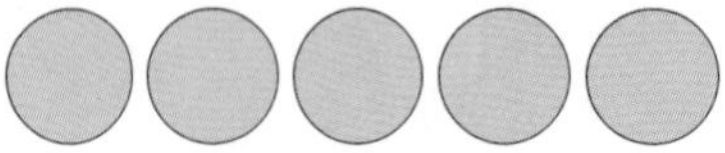

c. 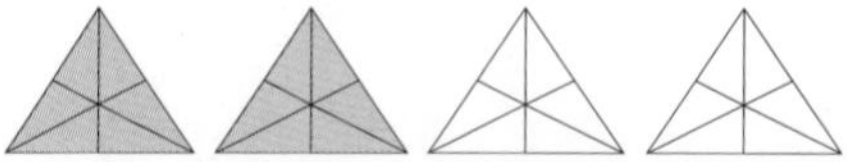

d. 

e. 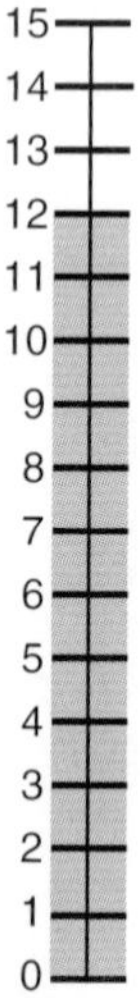


f. 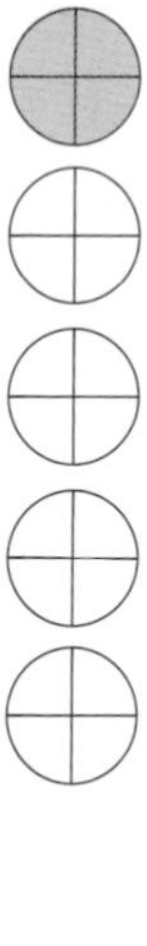

## Part 3

a.

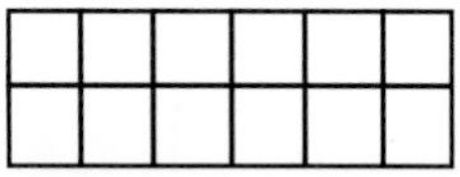

a. 2
x 6
[ ] squares

b.

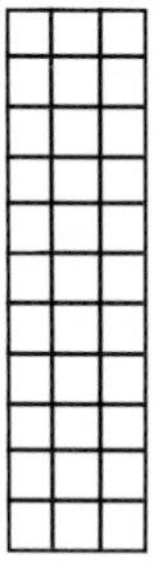

c.

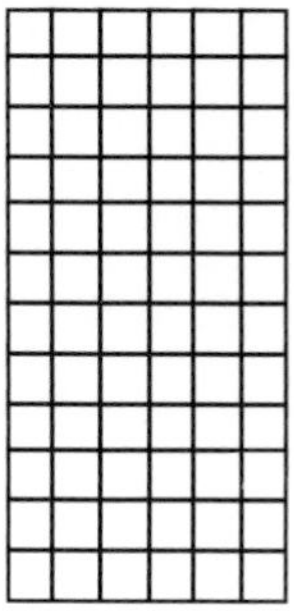

# Independent Work

## Part 4

**Write the column problem and answer for each family. Make a box around each answer.**

a. 812 ■ → 905

b. 62 418 → ■

c. 223 ■ ↓ 480

d. ■ 112 → 675

e. 23 ■ ↓ 480

## Part 5

**Copy each problem and work it.**

a. 764
x 5

b. 816
x 9

c. 427
x 2

**Part 6** **Write the fraction for each picture.**

a.

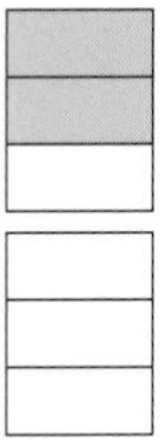

b.

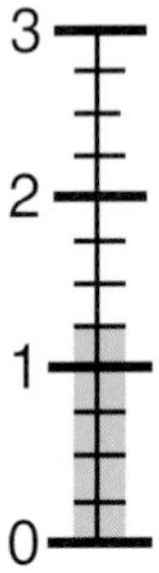


c.

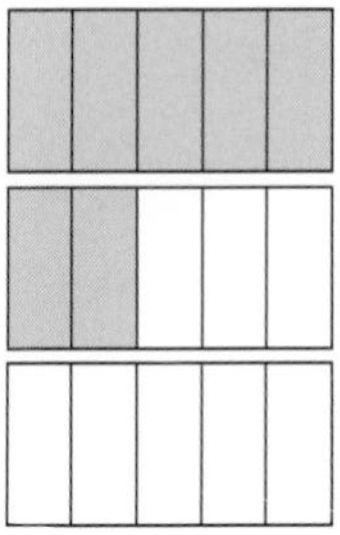

d.

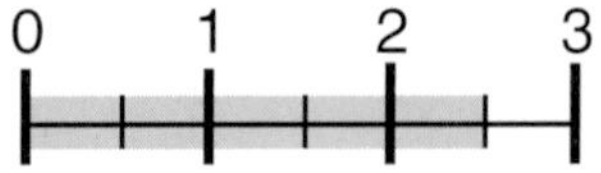

# Lesson 14

## Part 1

a. 5 + 3 ■ 8

b. 6 – ■ = 1

c. 2 ■ 8 = 16

d. ■ + 10 = 10

e. 8 ■ 4 = 4

f. 5 x ■ = 30

g. ■ – 10 = 50

h. 4 ■ 5 = 9

## Part 2

a. 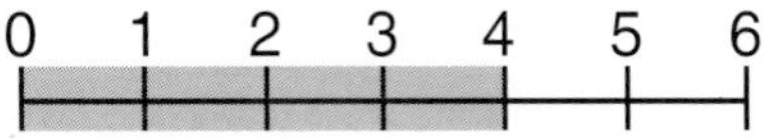


b. 

c. 

d. 

e. 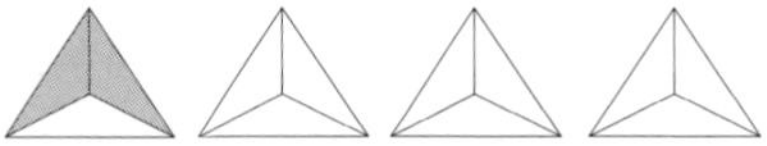

f. 

## Part 3

a. 943 – 38 = ■

b. 564 – 244 = ■

c. 614 – 367 = ■

## Part 4

- You've figured out the number of squares in rectangles.

- Some problems tell you that the units for each side are **feet.** Each square in this rectangle is one **square foot.**

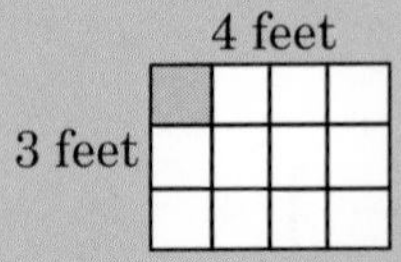

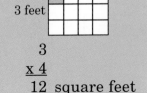

- So the name that goes in the answer is not just **squares,** but **square feet.**

$$\begin{array}{r} 3 \\ \underline{\times\ 4} \\ 12 \end{array}\ \text{square feet}$$

- Some problems tell you that the units for each side are **inches.** Each square in this rectangle is one **square inch.**

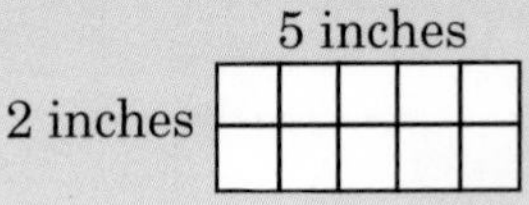


- Other rectangles have sides that are measured in **yards,** in **meters,** or even in **miles.**

$$\begin{array}{r} 2 \\ \underline{\times\ 5} \\ 10 \end{array}\ \text{square inches}$$

- You can use abbreviations for these units.

| | | |
|---|---|---|
| For the word | **square,** you write | **sq** |
| | **inches** | **in** |
| | **miles** | **mi** |
| | **feet** | **ft** |
| | **yards** | **yd** |

- Here's a rectangle without the squares shown:

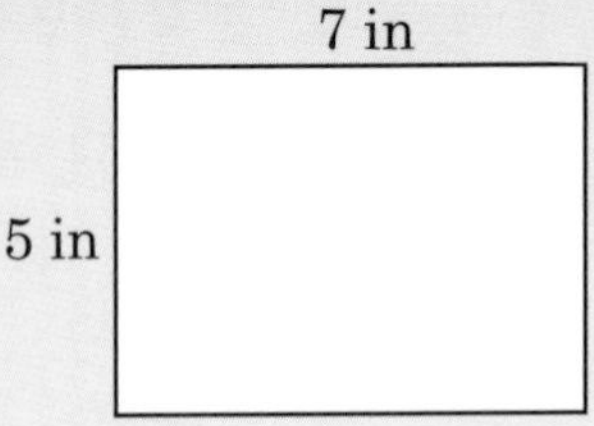


- The rectangle is 5 inches high. So there are 5 squares in each column. The rectangle is 7 inches wide. So there are 7 columns.

- Here's the multiplication problem and the answer for the area of the rectangle:

$$\begin{array}{r} \mathbf{5} \\ \underline{\mathbf{\times\ 7}} \\ \mathbf{35} \end{array}\ \textbf{sq in}$$

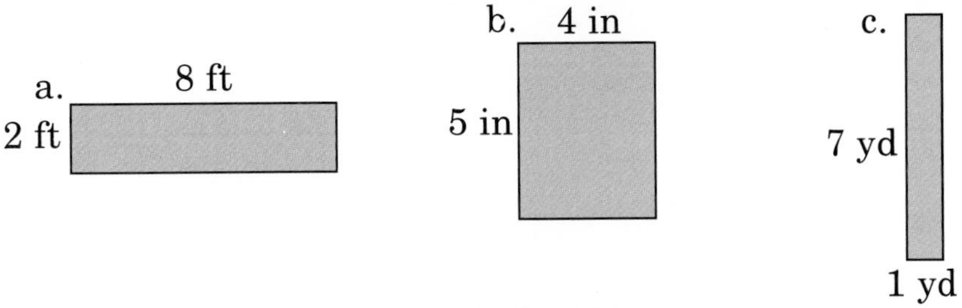


## Independent Work

### Part 5

**Write the column problem and answer for each family. Make a box around each answer.**

a. ■ 601 → 803

b. 461 38 → ■

c. 463 ■ → 471

### Part 6

**Write the symbols for each dollar-and-cents amount.**

a. 4 dollars and 6 cents

b. 91 dollars and 39 cents

c. 100 dollars and 40 cents

d. 18 dollars and no cents

### Part 7

**Copy each problem and work it.**

a. $\begin{array}{r} 37 \\ \underline{+\ \ 5} \end{array}$ b. $\begin{array}{r} 37 \\ \underline{\times\ \ 5} \end{array}$ c. $\begin{array}{r} 37 \\ \underline{-\ \ 3} \end{array}$ d. $\begin{array}{r} 370 \\ \underline{\times\ \ \ 5} \end{array}$ e. $\begin{array}{r} 370 \\ \underline{-\ \ 50} \end{array}$

### Part 8

**Write the answer to each problem.**

a. $\begin{array}{r} 5 \\ \underline{\times 4} \end{array}$ b. $\begin{array}{r} 6 \\ \underline{\times 1} \end{array}$ c. $\begin{array}{r} 4 \\ \underline{\times 2} \end{array}$ d. $\begin{array}{r} 5 \\ \underline{\times 8} \end{array}$ e. $\begin{array}{r} 6 \\ \underline{\times 9} \end{array}$ f. $\begin{array}{r} 9 \\ \underline{\times 3} \end{array}$ g. $\begin{array}{r} 5 \\ \underline{\times 5} \end{array}$

h. $\begin{array}{r} 0 \\ \underline{\times 3} \end{array}$ i. $\begin{array}{r} 7 \\ \underline{\times 5} \end{array}$ j. $\begin{array}{r} 8 \\ \underline{\times 9} \end{array}$ k. $\begin{array}{r} 6 \\ \underline{\times 5} \end{array}$ l. $\begin{array}{r} 2 \\ \underline{\times 8} \end{array}$ m. $\begin{array}{r} 5 \\ \underline{\times 9} \end{array}$ n. $\begin{array}{r} 9 \\ \underline{\times 7} \end{array}$

# Lesson 15

## Part 1

- Some students make terrible mistakes when they multiply by zero.

- Here's one mistake they make. They work the problem in the ones column properly.
- They carry the 2 to the tens column.
- Then they work the tens. They multiply 2 x 9. That's wrong. 103 x 9 **does not equal** 1087.

```
   1 2
   103
x    9
------
  1087
```

- Here's the problem worked the right way. The ones column is the same.
- The right way to work the tens is to multiply zero x 9. That's zero. Then add the 2 that is carried. The answer is 2 in the tens column.
- 103 x 9 **equals** 927.

```
    2
  103
x   9
-----
  927
```

- You won't get mixed up if you remember that zero times any number or any number times zero is zero.
- First you multiply. Then you add the number you carried.

## Part 2

| | | |
|---|---|---|
| a. ■ – 3 = 10 | b. 10 ■ 7 = 3 | c. 10 ■ 7 = 70 |
| d. 14 – ■ = 2 | e. ■ x 2 = 18 | f. 22 ■ 1 = 23 |

## Part 3

| | | |
|---|---|---|
| 3 | 6 | 9 |
| ___ | ___ | ___ |
| ___ | 24 | ___ |
| ___ | | |

| | | |
|---|---|---|
| 1 | 2 | 3 |
| 4 | 5 | 6 |
| 7 | 8 | 9 |
| 10 | | |

## Part 4

a. 2 x ■ = 2 b. 2 x 2 = ■ c. 3 x ■ = 9 d. 3 x 9 = ■

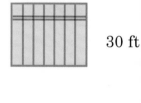

e. 2 x ■ = 6 f. 4 x 4 = ■ g. 2 x 8 = ■

## Part 5

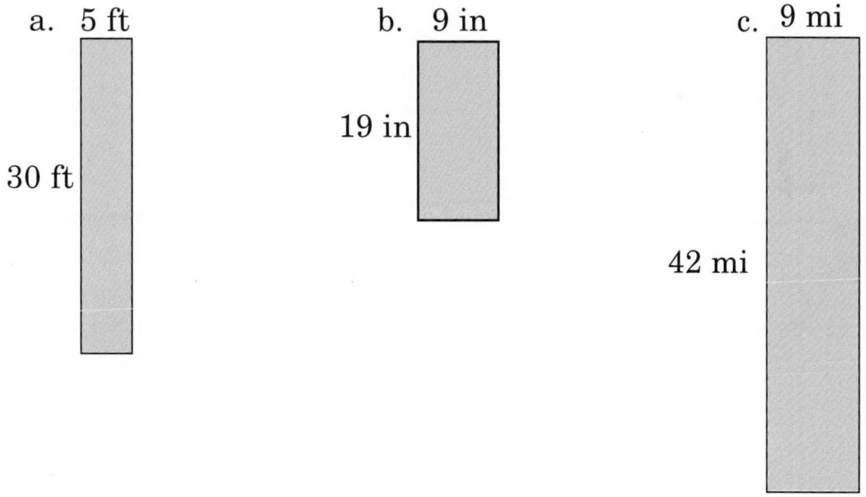


## Part 6

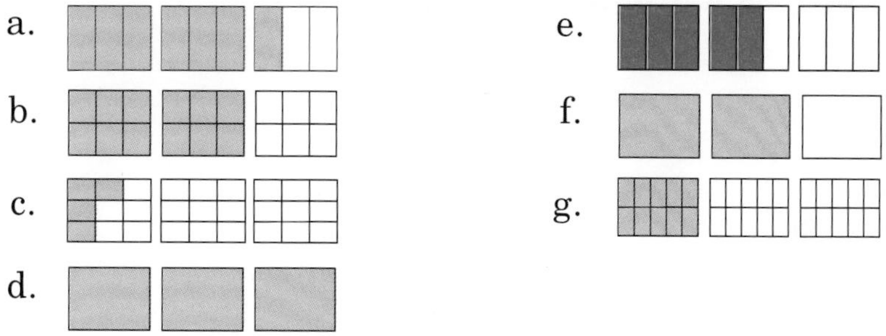

## Independent Work

### Part 7

**Write the addition problem or subtraction problem for each column. Make a box around each answer.**

| | | | |
|---|---|---|---|
| | | 82 | |
| | 90 | 29 | 119 |
| Total | 146 | | 257 |

### Part 8

**Write the time shown on each clock.**

a.

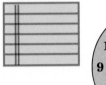

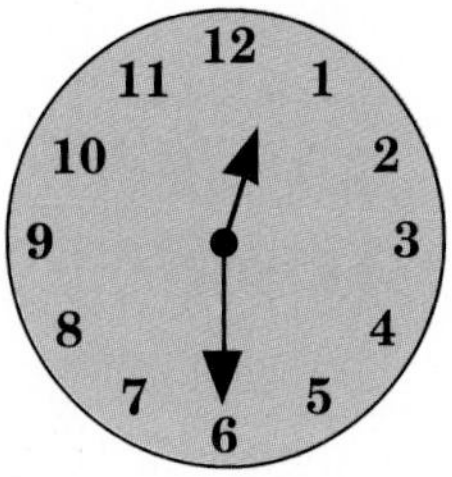

b.

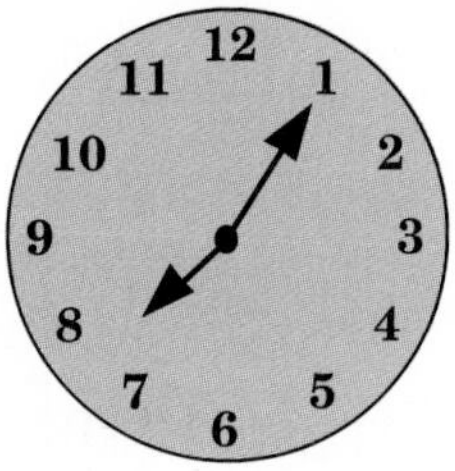

c.

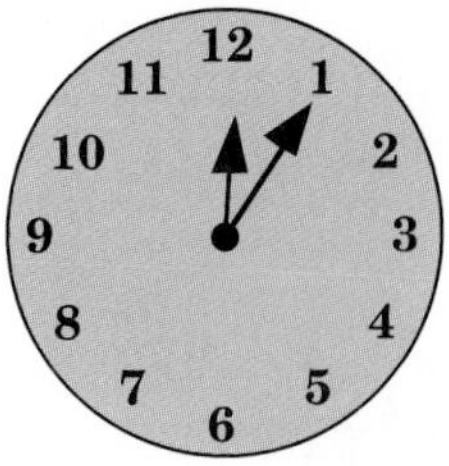

### Part 9

**Write the symbols for each dollar-and-cents amount.**

a. 1 dollar and 1 cent

b. 10 dollars and 10 cents

c. 100 dollars and no cents

d. 2 dollars and 22 cents

e. 19 dollars and 7 cents

### Part 10

**Copy each problem and work it.**

a. $\begin{array}{r} 83 \\ -\ \ 9 \\ \hline \end{array}$

b. $\begin{array}{r} 83 \\ +\ \ 9 \\ \hline \end{array}$

c. $\begin{array}{r} 83 \\ \times\ \ 9 \\ \hline \end{array}$

d. $\begin{array}{r} 803 \\ +\ \ \ 9 \\ \hline \end{array}$

e. $\begin{array}{r} 641 \\ -630 \\ \hline \end{array}$

f. $\begin{array}{r} 641 \\ +640 \\ \hline \end{array}$

**Part 11** **Write the answer to each problem.**

| | | | | |
|---|---|---|---|---|
| a. $5 \times 7$ | b. $9 \times 6$ | c. $3 \times 7$ | d. $2 \times 7$ | e. $2 \times 9$ |
| f. $5 \times 9$ | g. $2 \times 8$ | h. $9 \times 9$ | i. $1 \times 6$ | j. $5 \times 5$ |

24
27
We can't see your number. Turn your card around.

# Lesson 16

## Part 1

- For some problems, there are three names that go together just like the parts of a number family.
- Here are some names that go together:

  Small rabbits, big rabbits, all rabbits
  Dirty cars, cars that are not dirty, all cars
  Boys, girls, children
  Dead trees, living trees, all trees

- Here's a problem: **There were 8 new shoes. The rest were old shoes. There were 46 shoes in all. How many shoes were old shoes?**

- Here's the number family with the names:

  **new old all**
  ——————→
  **shoes**

- Here's the family with the numbers:

  new old all
  **8** ——————→ **46**
  shoes

- There were 38 old shoes.

  new old all
  8 **38** ——→ 46
  shoes

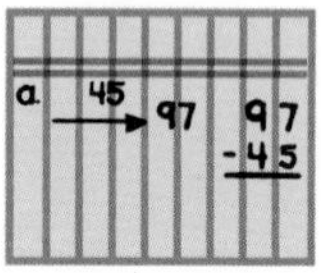


a. There were 61 black horses. There were 41 horses that were not black. How many horses were there in all?

b. There were 81 plants in all. 56 were alive. The rest were dead. How many plants were dead?

c. 45 bikes were broken. The rest were not broken. There were 278 bikes in all. How many were not broken?

## Part 2

a. $3 \times 6 = \blacksquare$  b. $3 \times \blacksquare = 6$  c. $2 \times \blacksquare = 10$

d. $2 \times 10 = \blacksquare$  e. $3 \times 9 = \blacksquare$  f. $3 \times \blacksquare = 9$

## Part 3

| | | |
|---|---|---|
| 3 | ___ | ___ |
| ___ | 15 | ___ |
| ___ | ___ | 27 |
| ___ | | |

| | | |
|---|---|---|
| 1 | 2 | 3 |
| 4 | 5 | 6 |
| 7 | 8 | 9 |
| 10 | | |

## Part 4

a. The picture has 4 parts shaded. There are 3 parts in each unit.

b. The picture has 30 parts in each unit. 14 parts are shaded.

c. The picture has 12 parts in each unit. 13 parts are shaded.

d. The numbers for the fraction are 11 and 12. The fraction is less than 1.

e. The numbers for the fraction are 3 and 7. The fraction is less than 1.

f. The picture has 3 parts in each unit. 4 parts are shaded.

g. The numbers for the fraction are 8 and 9. The fraction is more than 1.

## Part 5

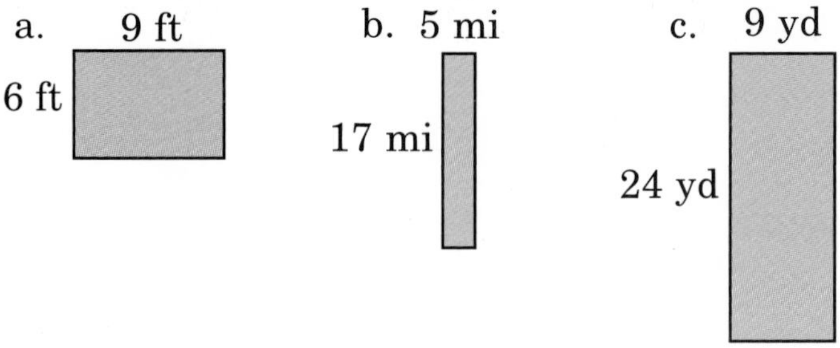


## Independent Work

## Part 6

**Write the fraction for each picture.**

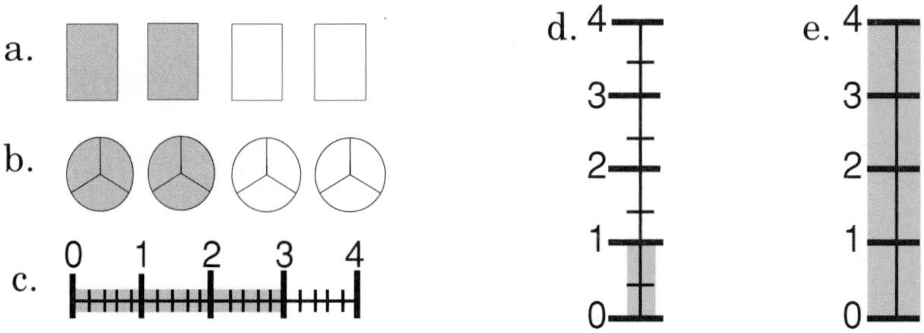


## Part 7

**Write each problem in a column and work it.**

a. 97 + 3 = ■

b. 97 x 3 = ■

c. 97 – 3 = ■

d. 197 – 98 = ■

## Part 8

**Write the column problem and the answer for each family. Make a box around each answer.**

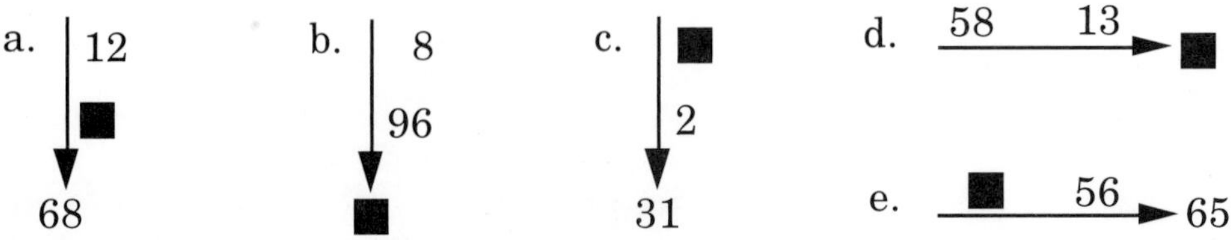

# Lesson 17

## Part 1

a. There were 89 children on the playground. 47 of the children were girls. How many children were boys?

b. 168 plants in a garden were alive. There were 209 plants in the garden. How many plants were dead?

c. 345 of the lights in a building were turned on. 236 were turned off. How many lights were there in all?

## Part 2

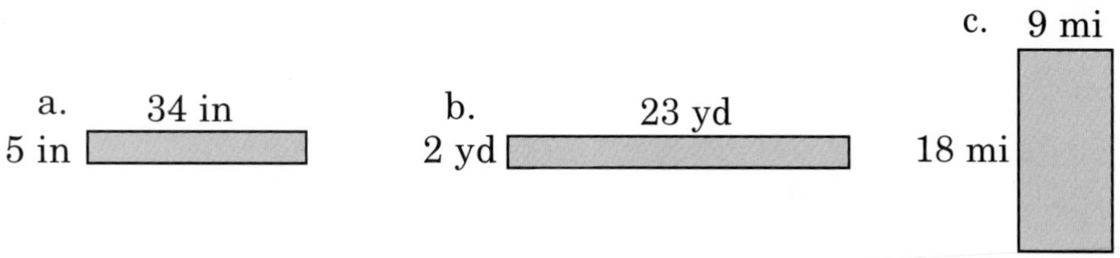


## Part 3

a. 784 – 752 = ■

b. 906 – 84 = ■

c. 114 – 68 = ■

d. 562 – 256 = ■

## Part 4

| | | |
|---|---|---|
| 1 | 2 | 3 |
| 4 | 5 | 6 |
| 7 | 8 | 9 |
| 10 | | |

## Part 5

- Some tables have totals at the end of the rows and totals at the bottom of the columns.
- These tables work just like number families that go along each row and number families that go down each column.

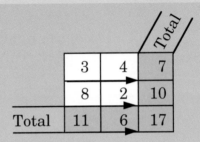

| | | | Total |
|---|---|---|---|
| | 3 | 4 | 7 |
| | 8 | 2 | 10 |
| Total | 11 | 6 | 17 |

- Here are the arrows for the rows:

  The first two numbers in each row are the small numbers. The total is the big number.

  If you add the small numbers for each row, you get the total for that row.

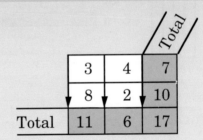

| | | | Total |
|---|---|---|---|
| | 3 | 4 | 7 |
| | 8 | 2 | 10 |
| Total | 11 | 6 | 17 |

- Here's the same table with the arrows for the columns:

  The top two numbers in each column are small numbers. The total at the bottom of each column is the big number.

  If you add the small numbers for each column, you get the total.

| | | | Total |
|---|---|---|---|
| | 3 | 4 | 7 |
| | 8 | 2 | 10 |
| Total | 11 | 6 | 17 |

- Here are the rules:

  **If there are two numbers in a row, you can figure out the missing number.**

  **If there's only one number in a row, you can't figure out the missing number.**

  **If there are two numbers in a column, you can figure out the missing number.**

  **If there's only one number in a column, you can't figure out the missing number.**

# Independent Work

## Part 6 Write the time shown on each clock.

a.

b.

c.

## Part 7 Write the answer to each problem.

a. $5 \times 8$

b. $5 \times 6$

c. $8 \times 2$

d. $8 \times 3$

e. $8 \times 0$

f. $8 \times 1$

g. $0 \times 3$

h. $7 \times 3$

i. $1 \times 3$

## Part 8 Write the fraction for each description. Circle each fraction that is more than 1.

a. The picture for fraction A shows 4 parts shaded. There are 7 parts in each unit.

b. Fraction B is more than 1. The numbers are 16 and 19.

c. The picture for fraction C shows 8 parts in each unit. 1 part is shaded.

d. Fraction D is less than 1. The numbers are 7 and 1.

e. The picture for fraction E shows 4 parts in each unit. 12 parts are shaded.

## Part 9

**Copy each problem and work it.**

a. 35 x 9

b. 176 x 3

c. 504 x 2

d. 409 x 9

## Part 10

**Write the column problem and answer for each family. Make a box around each answer.**

a. ■ 459 → 760

b. 888 349 → ■

c. 390 ■ ↓ 405

# Lesson 18

## Part 1

- Here's a table that has a total at the end of each row and a total at the bottom of each column.
- This table shows the number of hours two people worked on Monday and Tuesday.
- The heading for the first column is **Monday.** That column shows how many hours Dick worked on Monday and how many hours Mary worked on Monday.
- The number at the bottom of the column shows the total hours worked on Monday.

| | Monday | Tuesday | Total for both days |
|---|---|---|---|
| Dick | 4 | 6 | 10 |
| Mary | 2 | 8 | 10 |
| Total for both people | 6 | 14 | 20 |

- The second column shows the number of hours worked on **Tuesday.** The total for Tuesday was 14 hours.
- The **rows** show the hours for each person.
The top row is for Dick.
The next row is for Mary.

a. How many hours did Mary work on Tuesday?

b. How many hours did both people work on Monday?

c. How many hours did Dick work on both days?

d. How many hours did Mary work on both days?

e. How many hours did both people work on Tuesday?

## Part 2

a. 2 x 8 = ■ b. 2 x ■ = 8 c. 4 x ■ = 12

d. 3 x 8 = ■ e. 4 x 9 = ■ f. 4 x ■ = 8

## Part 3

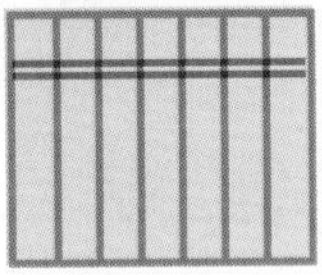

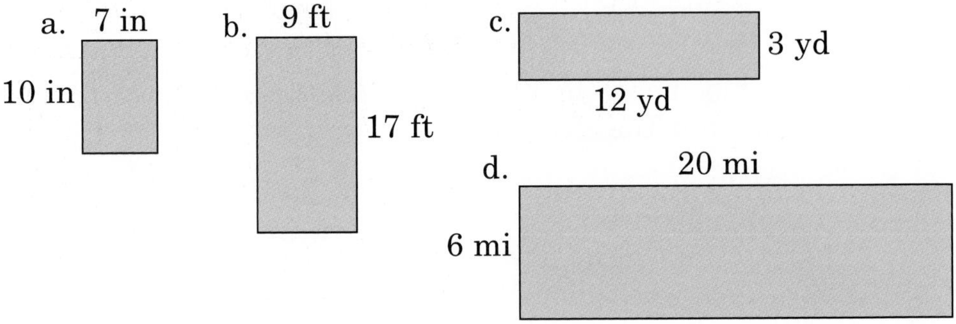


## Part 4

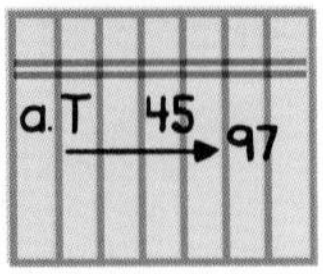


a. 400 students attended school on Tuesday. 71 of those students were late. The rest were on time. How many students were on time?

b. 50 cars on a lot need repairs. 211 cars on the lot do not need repairs. How many total cars are on the lot?

c. Some birds were in trees. 52 birds were not in trees. There were 199 birds in all. How many birds were in trees?

d. 45 of the vehicles were trucks. There were 570 vehicles in all. How many were not trucks?

## Part 5

| | | |
|---|---|---|
| 1 | 2 | 3 |
| 4 | 5 | 6 |
| 7 | 8 | 9 |
| 10 | | |

## Independent Work

### Part 6

**Write the fraction for each picture.**

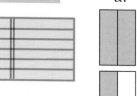

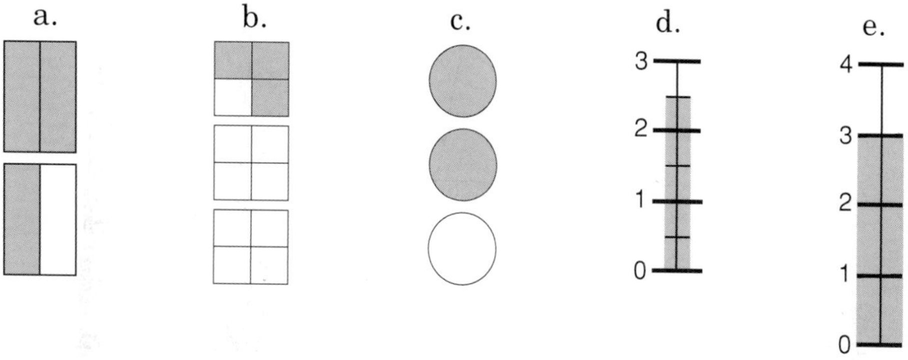


### Part 7

**Write the answer to each problem.**

a. 5 x 6 = ■  b. 3 x 9 = ■  c. 6 x 9 = ■

d. 4 x 3 = ■  e. 9 x 8 = ■  f. 4 x 2 = ■

### Part 8

**Write the symbols for each dollar-and-cents amount.**

a. 4 dollars and 13 cents  b. 13 dollars and 4 cents

c. 203 dollars and no cents  d. 1 dollar and 8 cents

### Part 9

**Write the column problem and the answer for each family. Make a box around each answer.**

a. 94, ■ → 386

b. ■, 321 → 517

c. 396, 107 → ■

**Part 10** **Copy each problem and work it.**

| a. | b. | c. | d. |
|---|---|---|---|
| 640 | 640 | 640 | 127 |
| x 5 | – 5 | + 50 | x 3 |

| e. | f. | g. | h. |
|---|---|---|---|
| 436 | 708 | 708 | 708 |
| – 139 | x 2 | + 2 | – 12 |

I have 123 worms. You have 4 more than I do. We'll figure out how many you have.

# Lesson 19

## Part 1

a. The floor of a room is 8 feet long and 5 feet wide. What is the area of the floor?

b. A board is 4 inches wide and 10 inches long. What is the area of the board?

c. A ranch is 25 miles long and 9 miles wide. How many square miles is the ranch?

## Part 2

This table shows the number of cars and trucks on lot A and lot B.

| | Lot A | Lot B | Total for both lots |
|---|---|---|---|
| Cars | 13 | 12 | 25 |
| Trucks | 16 | 14 | 30 |
| Total for both vehicles | 29 | 26 | 55 |

***Questions***

a. How many cars are on both lots?

b. How many vehicles are on lot B?

c. How many trucks are on lot B?

d. How many trucks are on both lots?

## Part 3

a. There were 71 yellow flowers in a garden. If there were 349 flowers in all, how many flowers were not yellow?

b. There were 135 boards on a wall. 29 of the boards were painted. The rest were not painted. How many boards were not painted?

## Part 4

| | | |
|---|---|---|
| 1 | 2 | 3 |
| 4 | 5 | 6 |
| 7 | 8 | 9 |
| 10 | | |

# Independent Work

## Part 5

**Write the time shown on each clock.**

a. b. c. d.

## Part 6

**Write the number problem and the answer for each family. Make a box around each answer.**

a. 236 632 → ■

b. 708 14 → ■

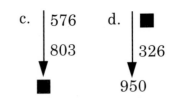

c. 576, 803 ↓ ■

d. ■, 326 ↓ 950

## Part 7

**Copy each problem and work it.**

a. $496 \times 9$

b. $496 - 9$

c. $496 + 9$

d. $496 - 90$

e. $496 + 90$

**Part 8** **For each problem, write the complete equation.**

a. 3 ■ 6 = 9 b. ■ – 5 = 1 c. ■ + 5 = 8

d. ■ x 5 = 15 e. 7 ■ 5 = 35

No, Alex, that is not the missing sign.
STOP
7 □ 5 = 35

# Lesson 20

## Part 1

a. $\frac{20}{5} =$

b. $\frac{9}{9} =$

c. $\frac{10}{2} =$

d. $\frac{18}{3} =$

e. $\frac{27}{3} =$

f. $\frac{14}{7} =$

# Lesson 21

## Part 1

a. $\frac{12}{4} =$

b. $\frac{24}{3} =$

c. $\frac{18}{2} =$

d. $\frac{18}{3} =$

## Part 2

a. A deck is 30 feet long and 9 feet wide. What is the area of the deck?

b. A sidewalk is 56 feet long and 3 feet wide. What is the area of the sidewalk?

c. A wall is 8 feet high and 12 feet wide. What is the area of the wall?

## Part 3

This table shows the number of trees that were planted in two parks.

| | Redwood trees | Hemlock trees | Total for both types |
|---|---|---|---|
| Hill Park | 73 | 16 | 89 |
| River Park | 42 | 65 | 107 |
| Total for both parks | 115 | 81 | 196 |

***Questions***

a. How many hemlocks were planted in River Park?

b. How many redwoods were planted in both parks?

c. How many total trees were planted?

d. How many hemlocks were planted in both parks?

e. How many redwoods were planted in Hill Park?

## Part 4

- Here's a multiplication problem with the middle number missing: $4 \times \square = 36$
- You can write it as a division problem.
- The 4 goes **in front** of the division sign. $4\overline{)\phantom{36}}$
- The box is **on top** of the division sign. $\begin{array}{r}\square\\4\overline{)\phantom{36}}\end{array}$
- The 36 goes **under** the division sign. $\begin{array}{r}\square\\4\overline{)36}\end{array}$
- The missing number in the multiplication problem is 9. That's the answer to the division problem. $\begin{array}{r}\boxed{9}\\4\overline{)36}\end{array}$
- Remember, the first value goes in front of the division sign. The next value goes on top. The number after the equal sign goes under the division sign.

## Independent Work

## Part 5

**Write the equation for each picture that shows a whole number. Remember to start your equation with the fraction.**

a. b. c. d. e. f.

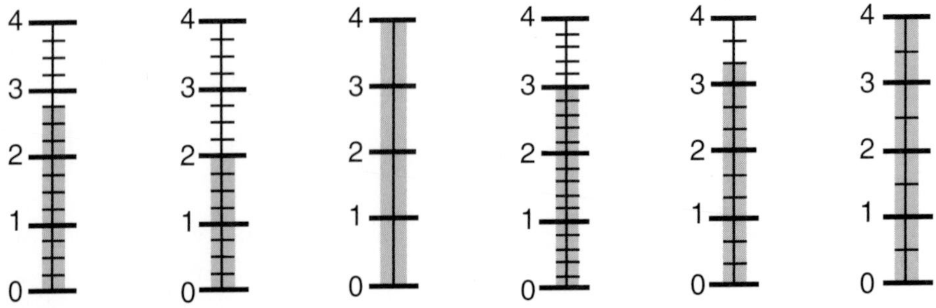

## Part 6

**Write the column problem and the answer for each family. Make a box around each answer.**

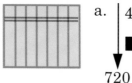

a. 48, ■ → 720

b. 48 720 → ■

c. ■ 72 → 480

## Part 7

**Copy each problem and work it.**

a. 370 x 3

b. 427 x 9

c. 305 x 6

d. 23 x 9

e. 35 + 860 + 12

f. 572 + 111 + 380

g. 435 + 15 + 255

h. 176 + 789 + 100

i. 374 − 86

j. 308 − 180

## Part 8

**Write the answer to each problem.**

| | | |
|---|---|---|
| a. 9 x 7 = ■ | e. 3 x 7 = ■ | i. 2 x 7 = ■ |
| b. 9 x 3 = ■ | f. 3 x 3 = ■ | j. 2 x 3 = ■ |
| c. 9 x 6 = ■ | g. 3 x 6 = ■ | k. 2 x 6 = ■ |
| d. 9 x 8 = ■ | h. 3 x 8 = ■ | l. 2 x 8 = ■ |

## Part J

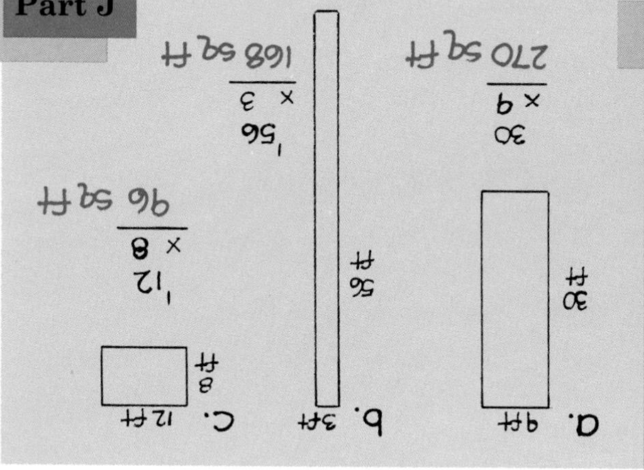


## Part K

a. 27 = 27
b. 207 > 27
c. 207 = 207
d. 14 < 41
e. 832 = 832
f. 1 > 0

# Lesson 22

## Part 1

a. A flag is 8 feet tall and 12 feet wide. How many square feet is the flag?

b. A floor is 6 yards wide and 10 yards long. How many square yards is the floor?

c. A shelf is 9 inches high and 24 inches wide. What is the area of the shelf?

## Part 2

a. $\frac{80}{8} =$ b. $\frac{36}{9} =$ c. $\frac{18}{3} =$ d. $\frac{18}{2} =$

## Part 3

a. There were 142 cars that were red. There were 256 cars that were not red. How many cars were there in all?

b. At the zoo, there were 461 animals in all. 87 of the animals were cats. How many of the animals were not cats?

## Part 4

- You can write fractions for whole numbers on a number line. These fractions are not for shaded parts. These fractions show what each whole number equals.
- The fraction at 1 equals the whole number 1.
- The fraction at 2 equals the whole number 2. And so forth.
- Here's a number line:

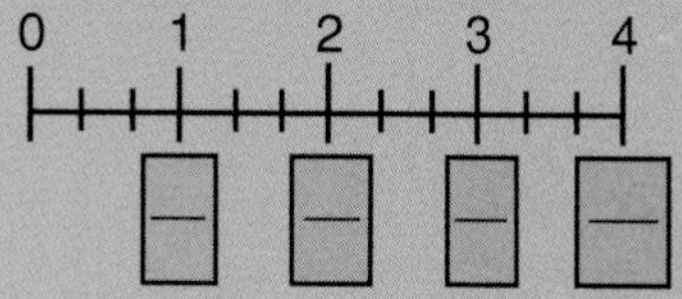


- Each unit is divided into the same number of parts. So the **bottom number** for each fraction **is the same.**
- There are 3 parts in each unit. So the bottom number of each fraction is 3.

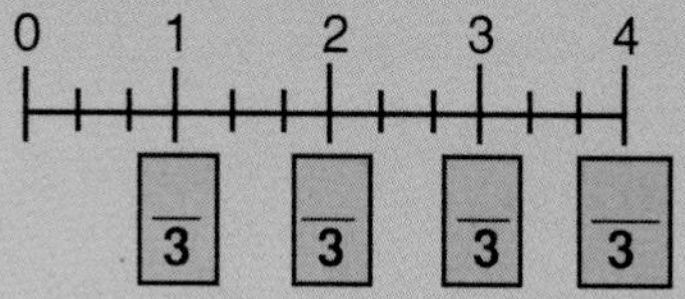


- The **top numbers** are **numbers for counting by 3.**

  The top number of the fraction at 1 equals 1 x 3.

  The top number that goes at 2 equals 2 x 3.

  The top number that goes at 3 equals 3 x 3.
- Here are the top numbers:

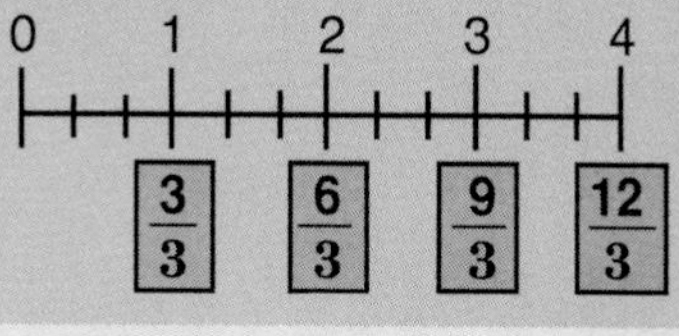


- Remember, the **bottom numbers** are the same for all the fractions. The **top numbers** are the numbers for counting by 3.
- And each fraction **equals the whole number above it.**

## Independent Work

### Part 5

**Copy each problem and work it.**

| | | | | |
|---|---|---|---|---|
| a. 560 x 3 | b. 903 x 9 | c. 567 x 2 | d. 305 x 8 | e. 401 x 2 |

### Part 6

**Write the column problem and the answer for each family. Make a box around each answer.**

a. 476 674 ──▶ ■

b. 209 ■ ──▶ 467 (vertical)

c. ■ 19 ──▶ 350

### Part 7

**Copy each problem and work it.**

| | | |
|---|---|---|
| a. 7 x 3 = ■ | e. 7 x 9 = ■ | i. 7 x 2 = ■ |
| b. 4 x 3 = ■ | f. 4 x 9 = ■ | j. 4 x 2 = ■ |
| c. 2 x 3 = ■ | g. 2 x 9 = ■ | k. 2 x 2 = ■ |
| d. 8 x 3 = ■ | h. 8 x 9 = ■ | l. 8 x 2 = ■ |

### Part J

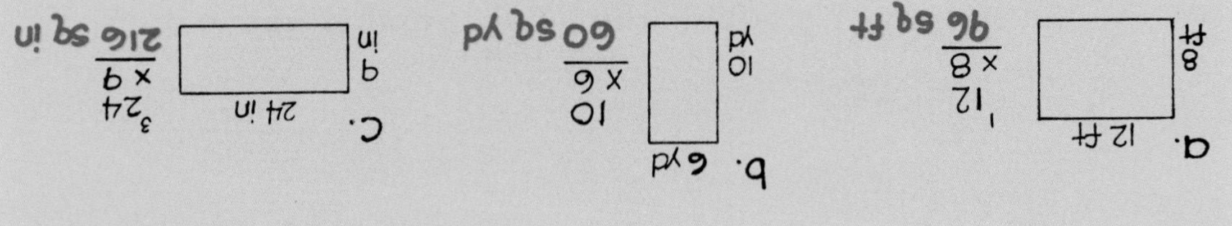

# Lesson 23

## Part 1

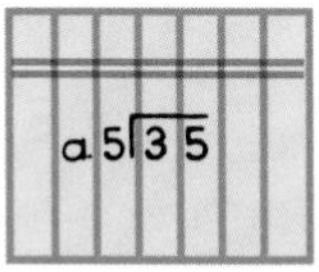


a. 5 x ■ = 15 b. 4 x ■ = 8 c. 7 x ■ = 7 d. 3 x ■ = 27

## Part 2

- It's easy to figure out the number you multiply by if the first number is 9.
- You look at the first digit of the number after the equal sign. The missing number is **1 more** than the first digit.
- Here's 9 x some number = 54:
  The first digit of 54 is 5.
  The missing number is 1 more than 5. It's 6.

  **9 x _ = 54**

  **9 x 6 = 54**
- Here's 9 x some number = 72:
  The first digit of 72 is 7.
  The missing number is 1 more than 7. It's 8.

  **9 x _ = 72**

  **9 x 8 = 72**
- Here's 9 x some number = 27:
  The missing number is 3.

  **9 x _ = 27**

  **9 x 3 = 27**

a. 9 x ■ = 63

b. 9 x ■ = 36

c. 9 x ■ = 81

d. 9 x ■ = 18

e. 9 x ■ = 54

f. 9 x ■ = 45

g. 9 x ■ = 72

## Part 3

a. $\frac{35}{5}$ b. $\frac{27}{5}$ c. $\frac{32}{5}$ d. $\frac{30}{5}$ e. $\frac{15}{5}$ f. $\frac{9}{5}$ g. $\frac{1}{5}$

h. $\frac{20}{5}$ i. $\frac{13}{5}$ j. $\frac{10}{5}$ k. $\frac{6}{5}$ l. $\frac{5}{5}$ m. $\frac{3}{5}$

## Part 4

This table shows the number of deer and squirrels that live in Hill Park and River Park.

| | Hill Park | River Park | Total for both parks |
|---|---|---|---|
| Deer | 23 | 40 | 63 |
| Squirrels | 19 | 86 | 105 |
| Total for both animals | 42 | 126 | 168 |

***Questions***

a. How many of both animals live in Hill Park?

b. How many deer live in River Park?

c. How many squirrels are there in both parks?

d. How many of both animals live in River Park?

e. How many squirrels live in Hill Park?

## Independent Work

## Part 5

**Write the answer to each problem.**

a. $3 \times 7$ b. $3 \times 8$ c. $3 \times 4$ d. $3 \times 6$ e. $3 \times 5$

f. $3 \times 9$ g. $2 \times 9$ h. $6 \times 9$ i. $8 \times 9$ j. $7 \times 9$

**Part 6** **For each problem, make a number family with names. Write the column problem and box the answer.**

a. There were cars on Elm Street. 34 of the cars were dirty. The rest were clean. There were 139 cars in all. How many were clean?

b. A farmer collected eggs. 150 were brown eggs. 489 were white eggs. How many eggs did the farmer collect in all?

c. In a garden, there were 42 blue flowers. The rest of the flowers were not blue. There were 150 flowers in all. How many were not blue?

**Part 7** **Write the column problem and the answer for each item.**

a. 273 + 80 + 109 = ■ b. 472 – 186 = ■ c. 635 + 557 = ■

d. 1743 + 92 + 846 = ■ e. 1056 – 829 = ■

No, write the problem in a column–not on a column.
1726 + 152 + 13 =

# Lesson 24

## Part 1

- When you add or subtract amounts for dollars and cents, you follow these rules:
- You show a dollar sign for the first number and you show a dollar sign in the answer.
- You make sure the decimal points are lined up.
- You write this:

```
$12.03
  1.76
+  .18
------
```

- You don't write this:

```
$12.03
 1.76
+.18
------
```

- To keep the decimal points lined up, turn your lined paper sideways and make your decimal points for the numbers. Make all the points on the same line. Then write your numbers.
- Then you add the numbers the way you always would, starting with the 1-cent column.

```
|$|1|2.|0|3|
| | |1.|7|6|
|+| | .|1|8|
|$| | .| | |
```

That's here: ↗

a. $9.24 – $.83 =

b. $8.75 + $20.03 + $6.09 =

## Part 2

This table shows the number of inches of snow that fell in January and February in two different parks.

| | January | February | Total for both months |
|---|---|---|---|
| Hill Park | 35 | 14 | 49 |
| River Park | 15 | 18 | 33 |
| Total for both parks | 50 | 32 | 82 |

***Questions***

a. Which park had more snowfall in February?

b. Which park had more snowfall in January?

c. In which month did River Park have less snowfall?

## Part 3

**Inches of Snowfall**

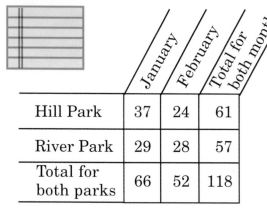

| | January | February | Total for both months |
|---|---|---|---|
| Hill Park | 37 | 24 | 61 |
| River Park | 29 | 28 | 57 |
| Total for both parks | 66 | 52 | 118 |

***Questions***

a. In which month did more snow fall in River Park?

b. In which month did less snow fall in Hill Park?

c. How many inches of snow fell in both parks in January?

d. How many inches of snow fell in Hill Park during both months?

e. Which park had more snowfall in February?

## Part 4

a. $\frac{11}{2}$ b. $\frac{6}{2}$ c. $\frac{18}{2}$ d. $\frac{5}{2}$ e. $\frac{1}{2}$ f. $\frac{8}{2}$ g. $\frac{3}{2}$

h. $\frac{16}{2}$ i. $\frac{20}{2}$ j. $\frac{9}{2}$ k. $\frac{10}{2}$

## Part 5

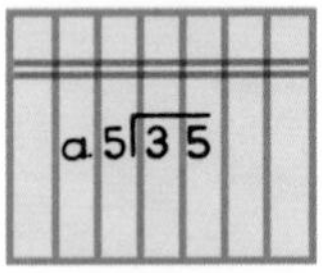


a. $6 \times \blacksquare = 18$  b. $6 \times \blacksquare = 6$  c. $5 \times \blacksquare = 30$

d. $2 \times \blacksquare = 10$  e. $9 \times \blacksquare = 45$  f. $1 \times \blacksquare = 7$

## Part 6

- If you know that one value is **more** than another value, you can put those values in a number family.
- Here's a sentence:

**14 is more than 11.**  11 ——→ 14

14 is more. It's the big number. 11 is one of the small numbers.

- If you know that one value is **less** than another value, you can put the values in a number family.
- Here's a sentence:

**24 is less than 30.**  24 ——→ 30

- You can do the same thing with names. You write the names above the number family.
- Here's a sentence:

**Pile M weighs more than pile R.**  R  M ——→

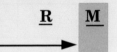

You write M above the space for the big number. You write R above the space for the second small number.

- Here's another sentence:

**Jan is shorter than Fran.**  Jan  Fran ——→

- Remember, the name for the bigger number is for the big number in the family. That's the value that is **farther, longer, bigger, heavier, older or faster.**

## Independent Work

**Part 7** **For each problem, make a number family with names. Write the column problem and box the answer.**

a. There were 89 boards. 19 boards were painted. The rest were not painted. How many boards were not painted?

b. Tim had 35 dimes. He had 127 coins that were not dimes. How many coins did he have in all?

c. 496 of the trees on a hill were oaks. The rest were not oaks. There were 611 trees on the hill. How many were not oaks?

d. There were perch and bass in a lake. There was a total of 680 fish. 135 were bass. How many were perch?

**Part 8** **Figure out the area of each rectangle. Write the multiplication problem with the 2-digit value on top. Write the units in the answer.**

a.
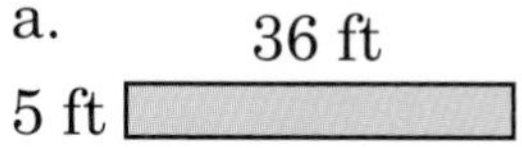
36 ft
5 ft

b.
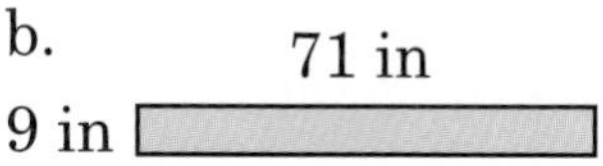
71 in
9 in

c.
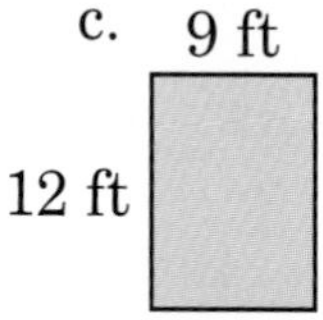
9 ft
12 ft

The area of this whole wall is 35 square feet.
And we've already painted a lot of square feet.

# Lesson 25

## Part 1

a. $\frac{7}{3}$ b. $\frac{36}{9}$ c. $\frac{35}{5}$ d. $\frac{20}{2}$ e. $\frac{20}{3}$

f. $\frac{20}{5}$ g. $\frac{20}{4}$ h. $\frac{20}{9}$ i. $\frac{20}{10}$

## Part 2

- Some word problems have sentences that compare two things. Here's a sentence that compares two things:

  **Ginger was (17 pounds) lighter than Ted.**

  17 pounds tells about the difference. Ginger is 17 pounds lighter. So the difference between Ginger and Ted is 17.

- You write 17 as the first number in the family. It's called the **difference** number.

  **difference**
  **17** →

- Then you read the sentence without the difference number:

  **Ginger was ( ) lighter than Ted.**

- That gives information about which is the name for the big number. Ginger was lighter. So Ted was heavier. That means Ted is the big number and Ginger is a small number.

  **difference Ginger Ted**
  **17** →

- When you work with sentences that compare, you can figure out the big number by reading the sentence without the difference number.

- Here's another sentence:

  **The cobra was (23 inches) longer than the green snake.**

  **difference**
  **23** →

- Here's the sentence without the difference number:

  **The cobra was ( ) longer than the green snake.**

  **difference** **green snake** **cobra**
  **23** →

- Here's another sentence with a difference number:

  **There were (56) fewer bugs than frogs.**

  **difference**
  **56** →

- Here's the sentence without the difference number:

  **There were ( ) fewer bugs than frogs.**

  **difference** **bugs** **frogs**
  **56** →

## Part 3

a. The dog was (15 pounds) lighter than the goat.

b. Pile R weighed (120 pounds) less than pile D.

c. The train traveled (230 miles) farther than the car.

d. The monkey was (14 inches) shorter than the chimp.

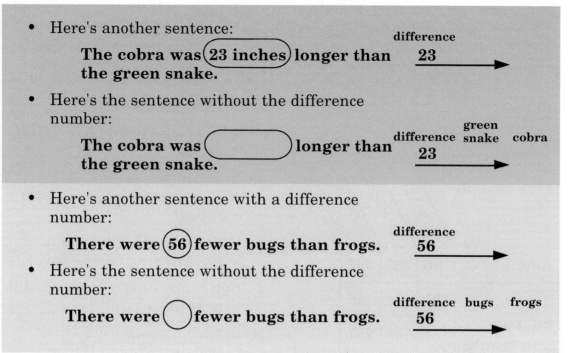

## Part 4

This table shows the number of rangers that were in Hill County and Donner County during July and August.

| | Hill County | Donner County | Total for both counties |
|---|---|---|---|
| July | 80 | 62 | 142 |
| August | 71 | 74 | 145 |
| Total for both months | 151 | 136 | 287 |

***Questions***

a. Which county had more rangers in August?

b. In which month were there more rangers working in both counties?

c. How many rangers were in Donner County during July?

d. How many rangers were in Hill County during July?

e. In which month were fewer rangers working in Donner County?

## Independent Work

## Part 5

**Write the column problem and the answer for each family. Make a box around each answer.**

a. 86 ■ → 107

b. 451, 17 ↓ ■

c. 51, ■ ↓ 235

## Part 6

**Copy each problem and work it.**

a. 186 x 9

b. 143 x 3

c. 9235 x 7

d. 409 x 5

## Part 7

**Figure out the area of each rectangle. Write the units in the answer.**

a.

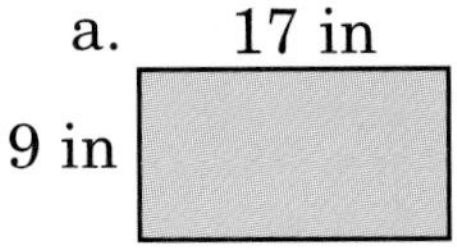


b.

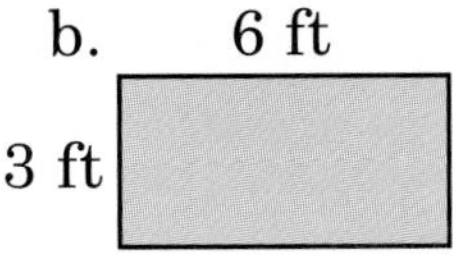


c.

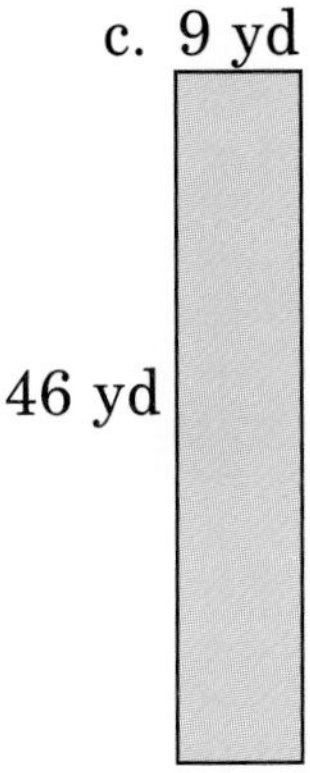


## Part 8

**Write the time shown on each clock.**

a.

b.

c.

## Part 9

**Write the fraction for each description.**

a. The fraction has numbers of 3 and 1. The fraction is less than 1.

b. The picture of the fraction shows 4 parts in each group. 12 parts are shaded.

c. The picture of the fraction shows 1 part in each group. 9 parts are shaded.

d. The fraction has numbers of 1 and 8. The fraction is more than 1.

e. The fraction has numbers of 15 and 16. The fraction is less than 1.

f. The fraction has numbers of 3 and 18. The fraction is more than 1.

# Lesson 26

## Part 1

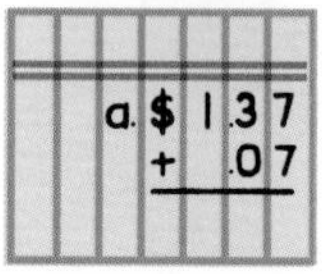


a. $17.14 – $9.06 =

b. $34.00 + $2.07 + $.92 =

c. $25.08 + $22.31 =

## Part 2

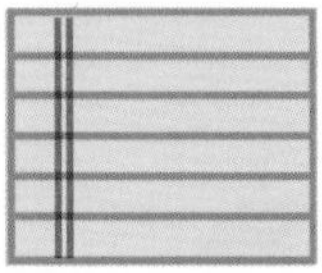

This table shows the number of frogs and fish that live in Lilly Lake and Star Lake.

| | Frogs | Fish | Total for both animals |
|---|---|---|---|
| Lilly Lake | 32 | 64 | 96 |
| Star Lake | 49 | 57 | 106 |
| Total for both lakes | 81 | 121 | 202 |

***Questions***

a. In which lake were there fewer fish?

b. Were there more frogs or more fish in Lilly Lake?

c. How many frogs and fish were in Lilly Lake?

d. Were there fewer frogs in Lilly Lake or in Star Lake?

e. How many frogs were in Star Lake?

## Part 3

a. The pole was (8 feet) shorter than the tree.

b. The bull is (307 pounds) heavier than the cow.

c. There are (68) fewer bottles than cans.

d. Jim ran (187 yards) farther than Steve ran.

e. A goat ate (328) fewer beans than a ram ate.

## Part 4

a. $\frac{4}{2}$  b. $\frac{16}{3}$  c. $\frac{16}{2}$  d. $\frac{16}{5}$  e. $\frac{18}{4}$

f. $\frac{18}{6}$  g. $\frac{18}{2}$  h. $\frac{15}{10}$  i. $\frac{10}{4}$  j. $\frac{25}{5}$

# Independent Work

## Part 5

**Copy each problem and work it. Use your calculator to check the answers.**

a. $\begin{array}{r} 17 \\ 381 \\ +\ 97 \\ \hline \end{array}$  b. $\begin{array}{r} 371 \\ \times\ 5 \\ \hline \end{array}$  c. $\begin{array}{r} 513 \\ -\ 106 \\ \hline \end{array}$  d. $\begin{array}{r} 123 \\ 156 \\ +\ 71 \\ \hline \end{array}$

e. $\begin{array}{r} 306 \\ \times\ 9 \\ \hline \end{array}$  f. $\begin{array}{r} 409 \\ \times\ 3 \\ \hline \end{array}$  g. $\begin{array}{r} 648 \\ -\ 56 \\ \hline \end{array}$  h. $\begin{array}{r} 20 \\ \times\ 2 \\ \hline \end{array}$

## Part 6

**Draw a diagram of each rectangle. Figure out the area. Write the units in the answer.**

a. A garden is 12 feet wide and 9 feet long. What is the area of the garden?

b. A wall is 8 feet high and 22 feet wide. What is the area of the wall?

c. A rectangle is 36 centimeters wide and 5 centimeters long. What is the area of the rectangle?

## Part 7

**Figure out the missing number in each family. Make a box around the answer.**

a.

b.
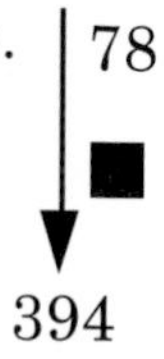


c.
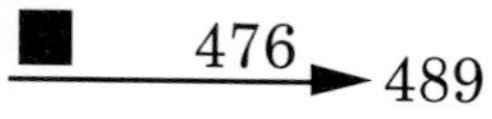

## Part 8 Write the answer to each problem.

a. 5 x 5 = ■
b. 2 x 7 = ■
c. 9 x 3 = ■
d. 8 x 0 = ■
e. 8 x 3 = ■
f. 8 x 2 = ■
g. 8 x 5 = ■
h. 8 x 1 = ■
i. 7 x 9 = ■
j. 7 x 3 = ■
k. 7 x 5 = ■
l. 7 x 1 = ■

When I told them to line up, I thought they'd do it in a row, not a column.

# Lesson 27

## Part 1

This table shows the number of meetings on the east coast and west coast during March and April.

| | East | West | Total for both coasts |
|---|---|---|---|
| March | 18 | 19 | 37 |
| April | 13 | 22 | 35 |
| Total for both months | 31 | 41 | 72 |

$$\begin{array}{r} 35 \\ -\ 22 \\ \hline 13 \end{array} \qquad \begin{array}{r} 31 \\ +\ 41 \\ \hline 72 \end{array}$$

$$\begin{array}{r} 31 \\ -\ 13 \\ \hline 18 \end{array} \qquad \begin{array}{r} 72 \\ -\ 35 \\ \hline 37 \end{array}$$

***Questions***

a. How many meetings were there on the east coast in April?

b. In which month were there more meetings on both coasts?

c. On which coast were there more meetings in March?

d. How many total meetings were there on both coasts during both months?

e. On which coast were there fewer meetings in March and April?

## Part 2

a. The snail is 47 inches shorter than the lizard.

b. Sue read 347 fewer words than Jill read.

c. The number of cans in lot M is 82 more than the number of cans in lot R.

d. The area of a bedroom is 108 square feet smaller than the area of a living room.

## Part 3

- When you add or subtract fractions, the bottom numbers must be the same. If the bottom numbers are not the same, the fractions don't tell about the same number line.
- Here's a problem: $\frac{3}{4} + \frac{5}{4} =$
- The bottom numbers are the same. The number line is divided into **fourths.** So the answer is **fourths:** $\frac{3}{4} + \frac{5}{4} = \frac{\ }{4}$
- You add the top numbers to see how many fourths you have. $\frac{3}{4} + \frac{5}{4} = \frac{8}{4}$
- Here's the problem on the number line for fourths:

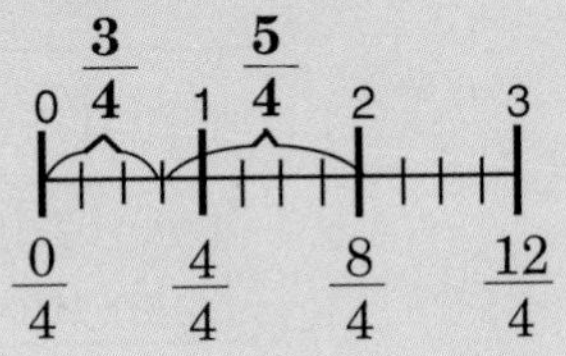


- **Remember, you add on top only. The bottom numbers tell about the units on the number line.**
- Here's a problem: $\frac{3}{4} + \frac{4}{3} =$
  You can't add these fractions the way they are written.
- The number line for $\frac{3}{4}$ has units divided into fourths.
- The number line for $\frac{4}{3}$ has units divided into thirds.
- The problem tells about two different number lines. So you can't work the problem the way it is written.

a. $\frac{3}{7} + \frac{2}{3} =$ b. $\frac{3}{9} + \frac{1}{9} =$ c. $\frac{5}{4} - \frac{4}{5} =$ d. $\frac{11}{3} - \frac{11}{9} =$

e. $\frac{7}{3} - \frac{3}{3} =$ f. $\frac{8}{7} + \frac{2}{7} =$ g. $\frac{8}{8} + \frac{8}{3} =$ h. $\frac{5}{6} - \frac{5}{6} =$

## Part 4

a. 7, 9, 2 b. 10, 5, 2 c. 4, 12, 3 d. 10, 6, 4

e. 7, 7, 1 f. 3, 15, 5 g. 3, 6, 3

# Independent Work

**Part 5** **Figure out the missing numbers. Make a box around each answer.**

a. 84 ■ → 960

b. 476 230 → ■

c. 692 410 → ■

d. ■ 296 → 851

**Part 6** **Copy each problem and work it.**

a. 371 x 9

b. 207 x 3

c. 804 x 5

d. 567 x 2

**Part 7** **Write each dollar-and-cent amount. Line up the decimal points.**

a. 36 dollars and 17 cents

b. 3 dollars and 2 cents

c. 180 dollars and no cents

d. 130 dollars and 9 cents

**Part 8** **Write each numeral.**

a. 56 thousand

b. 12 thousand 17

c. 8 thousand 1 hundred 92

d. 117 thousand 4

e. 312 thousand 2 hundred

## Part 9

**Make a number family for each problem. Write the column problem and box the answer.**

a. 13 of the tires were flat. The rest were not flat. There were 156 tires in all. How many were not flat?

b. There were 162 brown eggs and 809 white eggs. How many eggs were there in all?

c. 14 of the goats fainted. The rest did not faint. There were 90 goats. How many did not faint?

## Part 10

**Write the answer for each problem.**

| | | | | | |
|---|---|---|---|---|---|
| a. 3 x 6 | b. 9 x 3 | c. 9 x 2 | d. 3 x 7 | e. 2 x 3 | f. 4 x 3 |
| g. 5 x 4 | h. 3 x 3 | i. 3 x 8 | j. 9 x 4 | k. 5 x 3 | l. 3 x 1 |

## Part 11

**Write the fraction for each diagram. Circle the fractions that are more than 1.**

a. b. c. d.

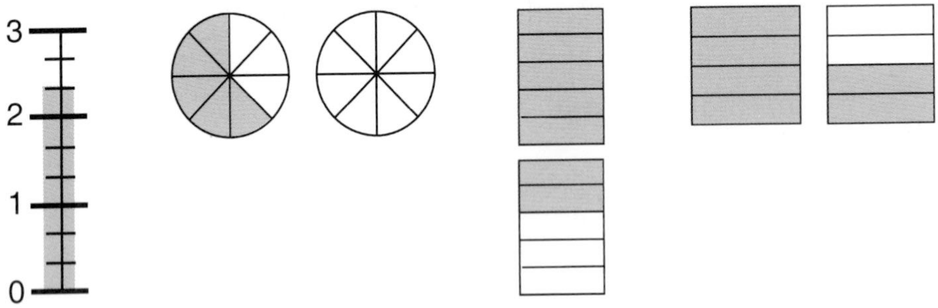

# Lesson 28

## Part 1

This table shows the number of red birds and blue birds on Elm Street and Maple Street.

| | Red birds | Blue birds | Total for both birds |
|---|---|---|---|
| Elm Street | 64 | 123 | 187 |
| Maple Street | 131 | 74 | 205 |
| Total for both streets | 195 | 197 | 392 |

$$\begin{array}{r} 131 \\ +\ 74 \\ \hline 205 \end{array} \qquad \begin{array}{r} 392 \\ -\ 197 \\ \hline 195 \end{array}$$

$$\begin{array}{r} 197 \\ -\ 74 \\ \hline 123 \end{array} \qquad \begin{array}{r} 392 \\ -\ 205 \\ \hline 187 \end{array}$$

***Questions***

a. On which street were there more red birds and blue birds?

b. Were there fewer blue birds or fewer red birds on Maple Street?

c. How many blue birds were there on Elm Street and Maple Street?

d. On which street were there more blue birds?

## Part 2

a. $\frac{7}{3} - \frac{5}{3} =$ b. $\frac{3}{5} + \frac{2}{3} =$ c. $\frac{4}{9} - \frac{4}{7} =$

d. $\frac{2}{9} + \frac{10}{9} =$ e. $\frac{12}{3} - \frac{6}{3} =$ f. $\frac{12}{12} - \frac{6}{6} =$

## Part 3

- To work word problems that compare two things, you start with the sentence that tells about the things compared. That sentence comes first in this problem:

  **The calf weighed 345 pounds less than the horse.**
  The horse weighed 509 pounds.
  How many pounds did the calf weigh?

- Here's the family for the first sentence: 345 is the difference number.

  **difference** **calf** **horse**
  **345** ⟶

- The next sentence in the problem gives a number for the horse:

  **The horse weighed 509 pounds.**

  **difference** **calf** **horse**
  **345** ⟶ **509**

- Now you can solve the problem because you have two numbers.
- Here's the subtraction problem and the answer: The calf weighed 164 pounds.

  **~~5~~ (4) [1]0 9**
  **− 3 4 5**
  **1 6 4 pounds**

a. Crow River is 259 miles shorter than Eagle River. Crow River is 416 miles long. How long is Eagle River?

b. The turtle is 47 years older than the mule. The turtle is 98 years old. How old is the mule?

c. There were 217 more green candies than red candies. There were 304 red candies. How many green candies were there?

d. A shark ate 570 fewer shrimp than a whale ate. The whale ate 742 shrimp. How many shrimp did the shark eat?

## Part 4

| | | | | |
|---|---|---|---|---|
| 4 | 8 | 12 | 16 | 20 |
| 24 | 28 | 32 | 36 | 40 |

| | | | | |
|---|---|---|---|---|
| 1 | 2 | 3 | 4 | 5 |
| 6 | 7 | 8 | 9 | 10 |

**Part 5** a. $\frac{5}{5}$ b. $\frac{5}{1}$ c. $\frac{8}{1}$ d. $\frac{8}{8}$ e. $\frac{12}{1}$

## Independent Work

**Part 6** Copy each fraction that equals a whole number. Write the equation to show the whole number it equals.

a. $\frac{15}{2}$ b. $\frac{15}{5}$ c. $\frac{15}{3}$ d. $\frac{15}{9}$ e. $\frac{15}{1}$

**Part 7** Write the fraction for each picture.

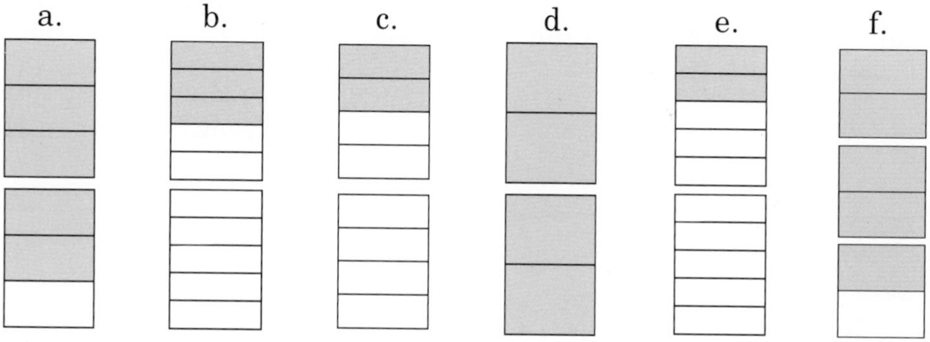


**Part 8** Figure out the missing number in each family. Make a box around the answer.

a. ■ 59 → 88  b. 478 ■ → 512  c. 71 961 → ■

**Part 9** Write the column problem and figure out the answer.

a. 407 x 3 = ■  b. 519 x 3 = ■

c. 6108 x 3 = ■  d. 4670 x 3 = ■

# Lesson 29

## Part 1

This table shows the tons of stones in two different parks.

| | River Park | Hill Park | Total for both parks |
|---|---|---|---|
| Green stones | 36 | 86 | 122 |
| Stones that are not green | 196 | 221 | 417 |
| Total stones | 232 | 307 | 539 |

***Questions***

a. Which park had more tons of green stones?

b. What was the total weight of all the stones in both parks?

c. How many tons of green stones were in River Park?

d. How many tons of stones that are not green were in both parks?

e. Were there fewer tons of stones in River Park or in Hill Park?

## Part 2

a. A ranch has 174 fewer sheep than cows. The ranch has 982 cows. How many sheep does the ranch have?

b. A goat ate 19 more carrots than a rabbit ate. The rabbit ate 7 carrots. How many carrots did the goat eat?

c. Jan weighs 108 pounds less than Dave. Jan weighs 113 pounds. How much does Dave weigh?

## Part 3

a. 7, 2, 14 b. 4, 8, 4 c. 15, 5, 10 d. 6, 2, 3 e. 0, 10, 10

## Part 4

a. $\frac{5}{1} - \frac{3}{1} =$ b. $\frac{6}{9} + \frac{6}{6} =$ c. $\frac{7}{4} + \frac{1}{4} =$

d. $\frac{10}{9} + \frac{8}{9} =$ e. $\frac{9}{10} + \frac{9}{8} =$ f. $\frac{15}{3} - \frac{4}{3} =$

## Part 5

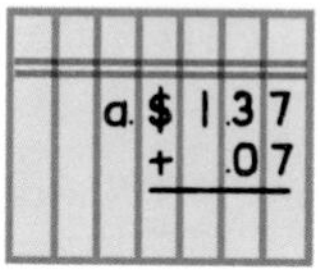

a. A person purchased the hat, the cat and the bat. How much did the person spend in all?

b. Joan bought a tire and a ball. How much did Joan spend?

# Independent Work

## Part 6

**Copy each problem and write the answer.**

a. $9\overline{)45}$ b. $2\overline{)16}$ c. $5\overline{)30}$ d. $3\overline{)15}$ e. $9\overline{)18}$

f. $2\overline{)8}$ g. $2\overline{)12}$ h. $5\overline{)25}$ i. $9\overline{)36}$

## Part 7

**Write the time shown on each clock.**

a.

b.

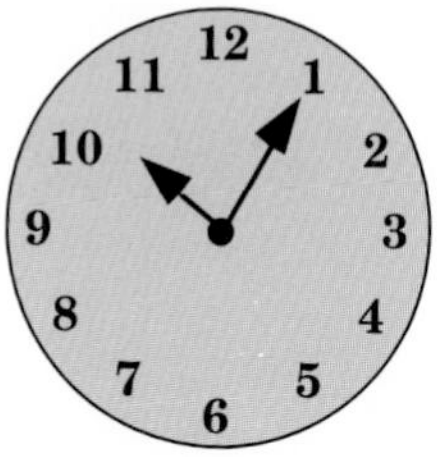

## Part 8

**Copy each equation. Figure out the missing sign or number.**

a. 4 ■ 3 = 12 b. 4 ■ 3 = 1 c. 4 ■ 3 = 7

d. 4 x ■ = 16 e. 4 x ■ = 0 f. 4 x ■ = 4

## Part 9

**Copy each problem and work it. Use your calculator to check the answers.**

a. $\begin{array}{r} 473 \\ +\ 1437 \\ \hline \end{array}$ b. $\begin{array}{r} 1567 \\ -\ 943 \\ \hline \end{array}$ c. $\begin{array}{r} 436 \\ \times\ 9 \\ \hline \end{array}$ d. $\begin{array}{r} 260 \\ 17 \\ +\ 91 \\ \hline \end{array}$

e. $\begin{array}{r} 216 \\ \times\ 5 \\ \hline \end{array}$ f. $\begin{array}{r} 478 \\ \times\ 3 \\ \hline \end{array}$ g. $\begin{array}{r} 708 \\ \times\ 9 \\ \hline \end{array}$

## Part 10

**Write each numeral.**

a. 46 thousand 6 b. 12 thousand 1 hundred 90

c. 4 thousand 2 d. 37 thousand 17

## Part 11

**Make a number family for each problem. Write the column problem and box the answer.**

a. Jim had 294 pieces of wood. He burned 185 pieces. How many pieces were not burned?

b. 56 cars were black. There were 560 cars in all. How many were not black?

## Part 12

**Figure the area of each rectangle. Write the units in the answer.**

a.

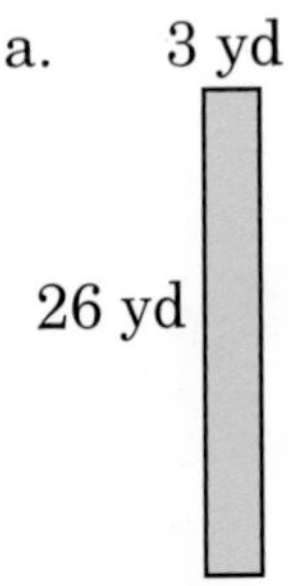


b.

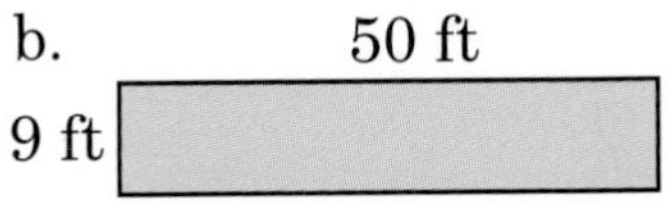

# Lesson 30

## Part 1

a. $5 = \frac{\quad}{1} = \frac{\quad}{3} = \frac{\quad}{2} = \frac{\quad}{4}$

b. $9 = \frac{\quad}{4} = \frac{\quad}{2} = \frac{\quad}{5} = \frac{\quad}{1}$

c. $4 = \frac{\quad}{3} = \frac{\quad}{1} = \frac{\quad}{2} = \frac{\quad}{5}$

## Part 2

a. $\frac{7}{3} - \frac{4}{3} =$

b. $\frac{9}{1} + \frac{5}{1} =$

c. $\frac{1}{5} - \frac{3}{4} =$

d. $\frac{7}{3} - \frac{7}{7} =$

e. $\frac{3}{4} + \frac{3}{7} =$

f. $\frac{8}{5} + \frac{7}{5} =$

g. $\frac{4}{5} - \frac{5}{11} =$

h. $\frac{2}{7} - \frac{1}{7} =$

# Lesson 31

## Part 1

| | | | | |
|---|---|---|---|---|
| 4 | | | 16 | |
| | | 32 | | 40 |

| | | | | |
|---|---|---|---|---|
| 1 | 2 | 3 | 4 | 5 |
| 6 | 7 | 8 | 9 | 10 |

## Part 2

a. $2\overline{)18}$  b. $9\overline{)18}$  c. $2\overline{)16}$  d. $9\overline{)72}$  e. $9\overline{)9}$

f. $9\overline{)45}$  g. $9\overline{)54}$  h. $1\overline{)7}$  i. $9\overline{)27}$

## Part 3

a. $5 = \frac{\quad}{10} = \frac{\quad}{1} = \frac{\quad}{6} = \frac{\quad}{4}$

b. $2 = \frac{\quad}{7} = \frac{\quad}{9} = \frac{\quad}{1} = \frac{\quad}{3}$

c. $9 = \frac{\quad}{1} = \frac{\quad}{5} = \frac{\quad}{3} = \frac{\quad}{10}$

## Part 4

a. Dandy Day Care purchased the doll and the phone. How much did Dandy Day Care spend in all?

b. Joan purchased the shoes, the book and the sunglasses. How much did Joan spend?

c. A little girl bought the shoes, the doll and the sunglasses. How much did the little girl spend in all?

## Part 5

a. 4, 3, 1　　b. 1, 8, 8　　c. 15, 1, 14　　d. 5, 15, 3

## Part 6

- Some problems divide by 5. You can figure out the answers to those problems by examining the number **under** the division sign.
- Here's what you do: **If the second digit of the number is zero, you multiply the first digit by 2. That's the answer.**
- Here's 40 divided by 5: The second digit of 40 is zero. So the answer is 2 times the first digit: **2 x 4 = 8**　　$\begin{array}{r} 8 \\ 5\overline{)40} \end{array}$
- Here's 30 divided by 5: The second digit is zero. So the answer is 2 times the first digit: **2 x 3 = 6**　　$\begin{array}{r} 6 \\ 5\overline{)30} \end{array}$
- Here's 20 divided by 5: The answer is 2 times the first digit: **2 x 2 = 4**　　$\begin{array}{r} 4 \\ 5\overline{)20} \end{array}$

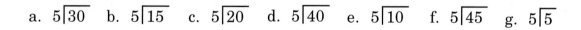

a. $5\overline{)30}$　b. $5\overline{)15}$　c. $5\overline{)20}$　d. $5\overline{)40}$　e. $5\overline{)10}$　f. $5\overline{)45}$　g. $5\overline{)5}$

## Part 7

a. A cat weighed 65 ounces more than a hamster. The cat weighed 93 ounces. How much did the hamster weigh?

b. Joe's garage pumped 487 fewer gallons of gas than Sally's station. Joe's garage pumped 632 gallons of gas. How many gallons of gas did Sally's station pump?

c. In a zoo, there were 401 more land animals than sea animals. There are 763 land animals. How many sea animals were there?

# Independent Work

## Part 8

**Copy each problem and work it.**

a. $\begin{array}{r} 608 \\ \underline{\times\ \ 3} \end{array}$ b. $\begin{array}{r} 704 \\ \underline{\times\ \ 3} \end{array}$ c. $\begin{array}{r} 407 \\ \underline{\times\ \ 9} \end{array}$ d. $\begin{array}{r} 608 \\ \underline{\times\ \ 9} \end{array}$

e. $\begin{array}{r} 502 \\ \underline{\times\ \ 9} \end{array}$ f. $\begin{array}{r} 301 \\ \underline{\times\ \ 9} \end{array}$ g. $\begin{array}{r} 103 \\ \underline{\times\ \ 3} \end{array}$

## Part 9

**Make a number family for each problem. Write the column problem and box the answer.**

a. 62 lights in a building were on. 147 lights were not on. How many lights are there in all?

b. There were 184 people on a plane. 77 of them did not wear hats. The rest wore hats. How many people wore hats?

## Part 10

**Copy each problem you can work and write the answer.**

a. $\frac{4}{3} - \frac{2}{3} =$ b. $\frac{7}{6} + \frac{7}{9} =$ c. $\frac{10}{2} + \frac{3}{4} =$

d. $\frac{12}{6} + \frac{8}{6} =$ e. $\frac{5}{8} - \frac{5}{8} =$ f. $\frac{15}{5} - \frac{5}{8} =$

## Part 11

**Write the fraction for each item.**

a. The numbers are 6 and 8. The fraction is more than 1.

b. The picture shows 11 parts in each unit. 7 parts are shaded.

c. The picture shows 3 parts shaded and 2 parts in each unit.

d. The numbers are 3 and 10. The fraction is less than 1.

# Lesson 32

## Part 1

a. $5\overline{)25}$  b. $5\overline{)35}$  c. $5\overline{)10}$  d. $5\overline{)20}$  e. $5\overline{)15}$

f. $5\overline{)45}$  g. $5\overline{)30}$  h. $5\overline{)5}$  i. $5\overline{)40}$

## Part 2

a. 0, 9, 9  b. 20, 10, 10  c. 14, 7, 2  d. 1, 1, 1

e. 35, 7, 5  f. 20, 1, 19  g. 20, 1, 20

## Part 3

| | | | | |
|---|---|---|---|---|
| 1 | 2 | 3 | 4 | 5 |
| 6 | 7 | 8 | 9 | 10 |

## Part 4

a. $3 = \frac{\phantom{0}}{4} = \frac{\phantom{0}}{10} = \frac{\phantom{0}}{5} = \frac{\phantom{0}}{1}$  b. $2 = \frac{\phantom{0}}{3} = \frac{\phantom{0}}{1} = \frac{\phantom{0}}{7} = \frac{\phantom{0}}{8}$

c. $5 = \frac{\phantom{0}}{8} = \frac{\phantom{0}}{2} = \frac{\phantom{0}}{7} = \frac{\phantom{0}}{3}$

## Part 5

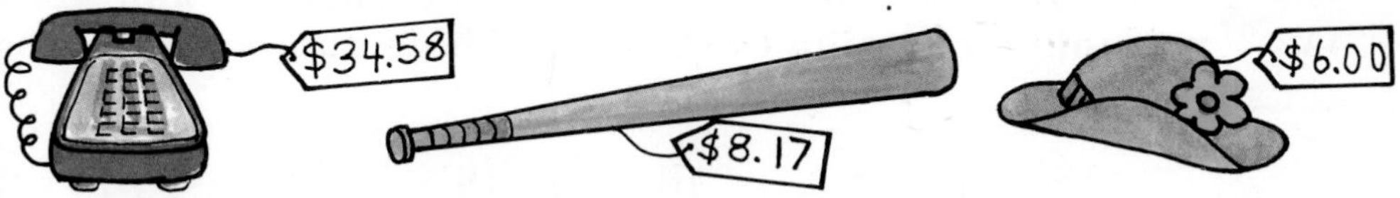


a. A person had $20.00. The person purchased the hat. How much did the person end up with?

b. Joe started out with $14.87. He bought the bat. How much did Joe end up with?

c. A club had $56.80 in the cash box. The club purchased the phone. How much did the club end up with?

## Part 6

a. $\frac{2}{5} + \frac{5}{2} =$  b. $\frac{12}{5} - \frac{9}{5} =$  c. $\frac{14}{9} - \frac{8}{9} =$

d. $\frac{2}{15} + \frac{24}{15} =$  e. $\frac{10}{5} - \frac{10}{4} =$  f. $\frac{10}{5} + \frac{9}{5} =$

# Independent Work

## Part 7

**Use the table in part 3 of your workbook to answer the questions.**

a. Were there more customers at Toy Palace or at Sports & Sports?

b. How many total people were at Sports & Sports?

c. In which store were there fewer workers?

d. How many customers were at Toy Palace?

## Part 8

**Write the division problem and the answer for each problem.**

a. $3 \times \blacksquare = 12$  d. $3 \times \blacksquare = 15$  g. $3 \times \blacksquare = 24$

b. $9 \times \blacksquare = 54$  e. $3 \times \blacksquare = 27$  h. $9 \times \blacksquare = 0$

c. $9 \times \blacksquare = 27$  f. $9 \times \blacksquare = 63$  i. $9 \times \blacksquare = 36$

## Part 9

**Copy the fractions that equal whole numbers and write the complete equation to show the whole number each fraction equals.**

a. $\frac{18}{3}$  b. $\frac{18}{4}$  c. $\frac{18}{9}$  d. $\frac{18}{2}$  e. $\frac{18}{5}$  f. $\frac{18}{10}$  g. $\frac{18}{1}$

**Part 10** **Make a number family for each problem. Write the column problem and box the answer.**

a. A truck was 59 feet long. A fence was 56 feet longer than the truck. How long was the fence?

b. A boy weighed 31 pounds more than a goat. The boy weighed 96 pounds. How many pounds did the goat weigh?

c. Cindy weighed 91 pounds less than her father. Her father weighed 193 pounds. How many pounds did Cindy weigh?

**Part 11** **Copy each problem and work it. Use your calculator to check the answers.**

a. $\begin{array}{r} 865 \\ -\ 378 \\ \hline \end{array}$

b. $\begin{array}{r} 203 \\ -\ 103 \\ \hline \end{array}$

c. $\begin{array}{r} 46 \\ 91 \\ +\ 124 \\ \hline \end{array}$

d. $\begin{array}{r} 274 \\ +\ 856 \\ \hline \end{array}$

**Part 12** **For each problem, draw the diagram and figure out the area.**

a. A sidewalk is 45 feet long and 5 feet wide. What is the area of the sidewalk?

b. A wall is 23 feet wide and 8 feet tall. How many square feet are in the wall?

c. A city is 13 miles long and 2 miles wide. How many square miles are in the city?

I see you've already painted more than one square foot.

# Lesson 33

## Part 1

a. A woman had $21.25. She bought diapers. How much did she end up with?

b. Jim had $5.10. He purchased milk. How much did he have left?

c. Susan needed detergent. She took $10.50 to the store and bought some detergent. How much money did Susan have left?

## Part 2

a. $4 = \frac{\quad}{7} = \frac{\quad}{5} = \frac{\quad}{9} = \frac{\quad}{3}$

b. $9 = \frac{\quad}{1} = \frac{\quad}{6} = \frac{\quad}{2} = \frac{\quad}{4}$

c. $1 = \frac{\quad}{1} = \frac{\quad}{8} = \frac{\quad}{10} = \frac{\quad}{47}$

## Part 3

a. $5\overline{)45}$   b. $9\overline{)45}$   c. $5\overline{)20}$   d. $5\overline{)30}$   e. $9\overline{)18}$

f. $5\overline{)15}$   g. $9\overline{)36}$   h. $5\overline{)35}$   i. $9\overline{)54}$   j. $5\overline{)40}$

k. $5\overline{)25}$   l. $9\overline{)81}$   m. $5\overline{)5}$   n. $9\overline{)27}$   o. $9\overline{)63}$

## Part 4

a. 3 x 20 = ■   b. 2 x 40 = ■   c. 3 x 90 = ■

d. 7 x 50 = ■   e. 4 x 90 = ■   f. 5 x 60 = ■

## Part 5

- You've learned to make number families for two different types of word problems.
- One type of word problem **compares two things** and tells about a difference number.
- Here's a sentence that **compares:**

  **The boat went 12 miles per hour faster than the raft.**

- That sentence lets you know that the number family has a **difference** number.
- Here's the family:
  12 is the **difference** number. The boat went faster, so **boat** is the name for the big number and **raft** is for a small number.

  **difference raft boat**
  **12** ——→

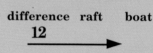

- Here's a sentence that **does not compare:**

  **There were 12 yellow rafts and the rest were blue.**

- For the number family, you write the names of the two things that are in the same class-yellow rafts and blue rafts.
- Here's the family for this sentence:
  The names for the small numbers are **yellow** and **blue.** The name for the big number is **all.** 12 is the number for yellow.

  **yellow blue all**
  **12** ——→
  **rafts**

a. 13 of the boats were dirty, and the rest were clean.

b. There were 14 more dirty boats than clean boats.

c. The bus held 51 more people than the van held.

d. There were boys and 18 girls in the park.

e. Some of the lights were turned on, and 19 were turned off.

f. Rita was 26 years younger than her mother.

## Independent Work

### Part 6

**The table in part 2 of your workbook shows the number of diamonds and rubies in two different mines. Use the table to answer these questions.**

a. Were there more rubies or diamonds found in Al's mine?

b. How many total gems were found in both mines?

c. Which mine had fewer diamonds?

d. How many total rubies were there?

### Part 7

**Copy each problem you can work and write the answer.**

a. $\frac{13}{3} - \frac{4}{3} =$ b. $\frac{9}{14} + \frac{5}{14} =$ c. $\frac{11}{5} - \frac{3}{4} =$

d. $\frac{7}{3} - \frac{7}{7} =$ e. $\frac{13}{4} + \frac{3}{7} =$ f. $\frac{18}{5} + \frac{2}{5} =$

### Part 8

**Copy each problem and write the answer.**

a. $9\overline{)27}$ d. $5\overline{)15}$ g. $3\overline{)21}$ j. $9\overline{)18}$

b. $3\overline{)27}$ e. $3\overline{)6}$ h. $3\overline{)12}$ k. $3\overline{)18}$

c. $3\overline{)15}$ f. $2\overline{)6}$ i. $3\overline{)24}$ l. $2\overline{)18}$

### Part 9

**Copy each problem and write a complete equation.**

a. 3 x ■ = 6 b. 3 ■ 3 = 6 c. 3 + 12 = ■

d. 3 ■ 2 = 1 e. 3 x ■ = 12

**Part 10** **Write the equation to show the fraction and the whole number each fraction equals.**

a. $\frac{81}{9}$ b. $\frac{35}{5}$ c. $\frac{6}{1}$ d. $\frac{20}{4}$ e. $\frac{10}{10}$

**Part 11** **Make a number family for each problem. Write the column problem and box the answer.**

a. A cow weighs 304 pounds less than a bull. The bull weighs 2340 pounds. How many pounds does the cow weigh?

b. Donna ate 31 fewer peanuts than Jerry ate. Donna ate 21 peanuts. How many peanuts did Jerry eat?

**Part 12** **Copy each problem and work it. Use your calculator to check the answers.**

a. 471 x 4 = ■ b. 506 x 9 = ■ c. 3876 x 4 = ■

3 6 9
12 15 18
21 24 27
30
Alex, you're not using the right map.
3⟌27

# Lesson 34

**Part 1**

a. $4 = \frac{\phantom{0}}{7} = \frac{\phantom{0}}{1} = \frac{\phantom{0}}{5}$

b. $\frac{6}{3} =$

c. $\frac{7}{3} - \frac{2}{3} =$

d. $\frac{3}{4} + \frac{10}{4} =$

e.

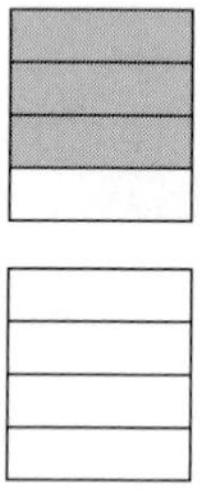

f. The numbers are 8 and 5. The fraction is less than 1.

g. The picture shows 7 parts shaded and 6 parts in each unit.

**Part 2**

| | | |
|---|---|---|
| 1 | 2 | 3 |
| 4 | 5 | 6 |
| 7 | 8 | 9 |
| 10 | | |

a. $3\overline{)21}$  b. $3\overline{)18}$  c. $3\overline{)6}$  d. $3\overline{)12}$  e. $3\overline{)24}$  f. $3\overline{)15}$

## Part 3

- You're going to do work with mixed numbers. A mixed number is a combination of a whole number plus a fraction.
- Sometimes mixed numbers are written like this: $2\frac{1}{3}$
- You'll work with mixed numbers that are written like this: $2 + \frac{1}{3}$
- These are mixed numbers:

A. $3 + \frac{1}{5}$ B. $4 + \frac{2}{7}$

The fraction part of the mixed number tells about the number line for the mixed number.

Mixed number A goes on a number line that shows **fifths.**

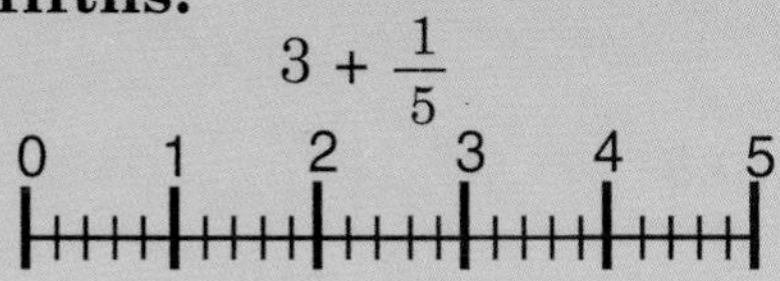


Mixed number B goes on a number line that shows **sevenths.**

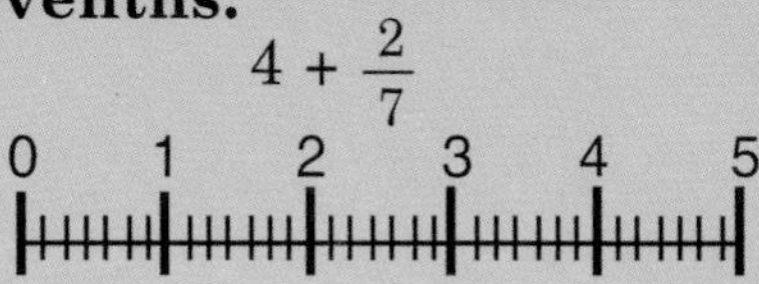


- To show the mixed number on a number line, you go to the marker for the whole number. Then you count parts for the fraction.

For $3 + \frac{1}{5}$, you go to the whole number 3. Then you count 1 more part. That's $\frac{1}{5}$.

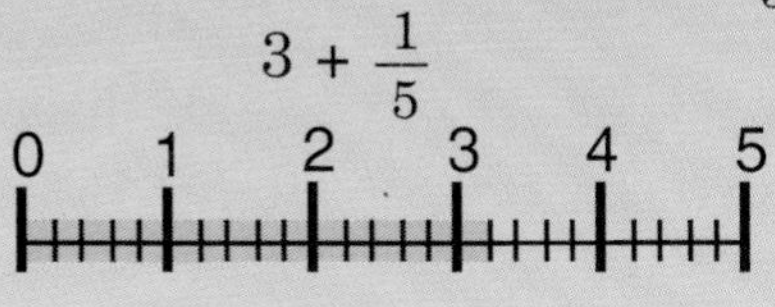


For $4 + \frac{2}{7}$, you go to the whole number 4. Then you count 2 more parts. That's $\frac{2}{7}$.

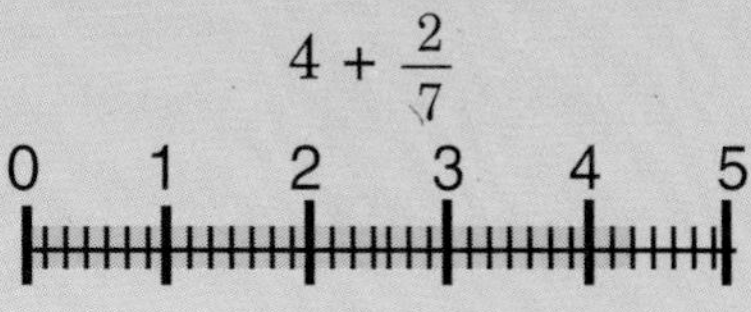


- Remember, mixed numbers are whole numbers **plus fractions.**

## Part 4

a. There were 34 more flowers than other plants.

b. There were 14 red cars on the lot, and the rest were not red.

c. Some doors in the building were open, and 72 doors were closed.

d. Fran was 13 centimeters shorter than her mother.

# Independent Work

## Part 5

**Make a number family for each problem. Write the column problem and box the answer.**

a. There were 145 posts. 83 of them had termites in them. The rest did not have termites. How many posts did not have termites?

b. A year has 365 days. In Suntown, 134 days were cloudy. The rest were sunny. How many days were sunny in Suntown?

c. A bag had marbles in it. 45 of the marbles were red. 112 marbles were not red. How many marbles were there in all?

## Part 6

**Copy each problem and work it.**

a. $870 \times 3$

b. $495 \times 3$

c. $1084 \times 3$

d. $3506 \times 9$

e. $4150 \times 9$

## Part 7

**Write the answer to each problem.**

a. 4 x 5 = ■

b. 6 x 4 = ■

c. 4 x 3 = ■

d. 4 x 8 = ■

e. 2 x 4 = ■

f. 7 x 4 = ■

g. 4 x 6 = ■

h. 1 x 4 = ■

i. 4 x 9 = ■

j. 4 x 7 = ■

k. 0 x 4 = ■

l. 8 x 4 = ■

## Part J

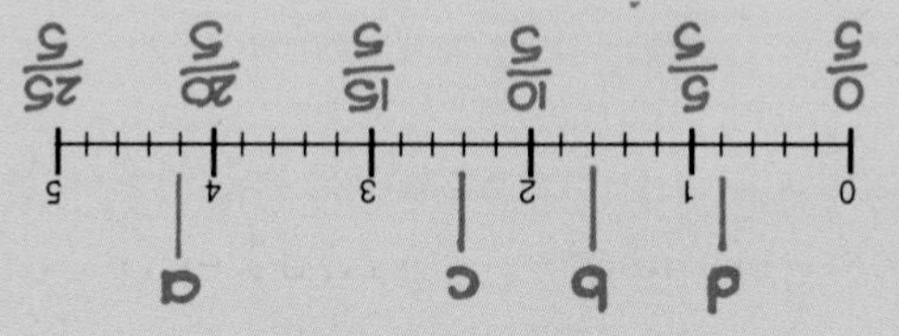

# Lesson 35

## Part 1

a. There were 7 hungry children, and the rest were not hungry.

b. The train was 117 yards longer than the truck.

c. In March, some of the days were cloudy, and 19 days were sunny.

d. In March, there were 11 more rainy days than sunny days.

## Part 2

a. $8 = \frac{\quad}{3}$   b. $4 = \frac{\quad}{5}$   c. $9 = \frac{\quad}{2}$   d. $6 = \frac{\quad}{3}$   e. $6 = \frac{\quad}{5}$

## Part 3

| | | |
|---|---|---|
| 1 | 2 | 3 |
| 4 | 5 | 6 |
| 7 | 8 | 9 |
| 10 | | |

a. $3\overline{)18}$   b. $3\overline{)9}$   c. $3\overline{)15}$   d. $3\overline{)6}$

e. $3\overline{)21}$   f. $3\overline{)3}$   g. $3\overline{)24}$   h. $3\overline{)12}$

## Independent Work

## Part 4

**Use the table in part 4 of your workbook to answer the questions.**

a. Were there more fleas at Al's Pet Shop or Tim's Kennel?

b. How many fleas were in Al's Pet Shop?

c. How many cat fleas were there in both places?

d. Were there fewer cat fleas or dog fleas in Al's Pet Shop?

## Part 5

**Write the fractions for a, b and c. Write complete equations for d through h.**

a. The numbers of a fraction are 3 and 1. The fraction is more than 1.

b. The picture of a fraction shows 6 parts shaded and 2 parts in each unit.

c.

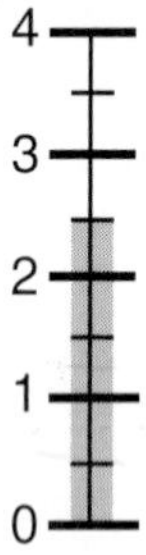


d. $5 = \frac{\phantom{0}}{1} = \frac{\phantom{0}}{10} = \frac{\phantom{0}}{9}$

e. $\frac{18}{3} =$

f. $\frac{45}{9} =$

g. $\frac{12}{5} - \frac{8}{5} =$

h. $\frac{17}{20} + \frac{2}{20} =$

## Part 6

**Write the answer to each problem.**

| | | |
|---|---|---|
| a. 4 x 8 = ■ | d. 4 x 9 = ■ | g. 4 x 1 = ■ |
| b. 4 x 2 = ■ | e. 4 x 7 = ■ | h. 4 x 4 = ■ |
| c. 4 x 0 = ■ | f. 4 x 5 = ■ | i. 4 x 6 = ■ |

## Part 7

**Use your calculator to check the answer to each problem. Copy the letter of each problem that has a wrong answer. Write the correct answer. Don't do anything to problems that have the correct answer.**

```
a.    1 1        b.   615          c.    2 2        d.     1
      4 8 1          7̸6̸¹4                8 8          5 5 8
        7 2        − 4 6 5                9 9        +   3 2
    + 9 1 1          2 9 9            + 3 3 3          5 0 0
    1 4 7 4                             5 2 0
```

## Part 8

**Copy each problem and work it.**

```
a.  379      b.  379      c.  379      d.  608
  x   2        x   4        x   3        x   9
```

## Part 9

**Make a number family. Write the column problem and box the answer.**

a. Andy's shoes had 461 more grains of sand in them than Cindy's shoes. There were 891 grains of sand in Andy's shoes. How many grains of sand were in Cindy's shoes?

b. There are 58 fewer trucks than cars on Elm Street. There are 160 trucks. How many cars are on Elm Street?

## Part 10

**Write the addition problem for each item and figure out how much each person spent.**

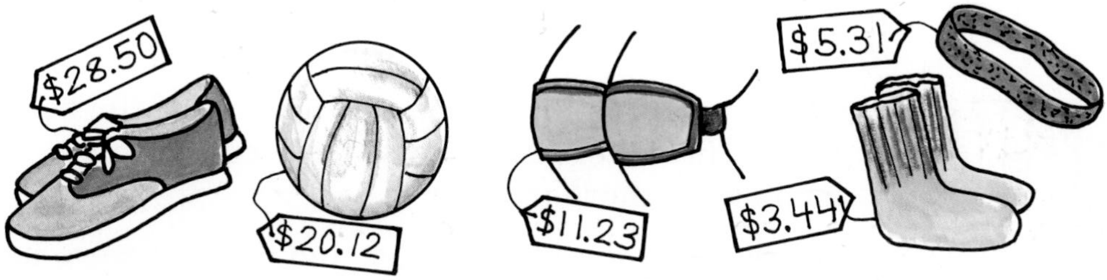


a. Juan bought the headband, the volleyball and the socks. How much did he spend?

b. Marsha bought the shoes and the kneepads. How much did she spend?

## Part J

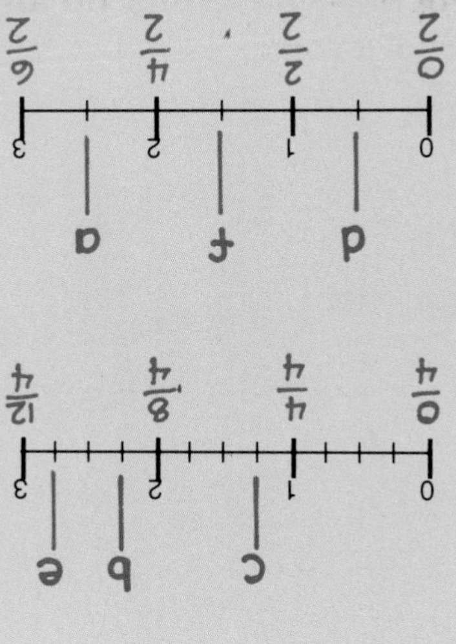

# Lesson 36

## Part 1

| | | | | |
|---|---|---|---|---|
| 4 | 8 | 12 | 16 | 20 |
| 24 | 28 | 32 | 36 | 40 |

| | | | | |
|---|---|---|---|---|
| 1 | 2 | 3 | 4 | 5 |
| 6 | 7 | 8 | 9 | 10 |

## Part 2

a. There were 14 more blue boats than white boats. There were 89 blue boats. How many white boats were there?

b. There were 14 blue boats, and the rest of the boats were not blue. There were 109 boats that were not blue. How many boats were there in all?

c. There were small umbrellas, and the rest were not small. 68 umbrellas were not small. There were 271 umbrellas in all. How many were small?

d. There were 120 fewer hungry dogs than full dogs. 134 dogs were full. How many were hungry?

## Part 3

a. $3 = \frac{\quad}{6}$  b. $4 = \frac{\quad}{2}$  c. $5 = \frac{\quad}{10}$  d. $2 = \frac{\quad}{8}$  e. $9 = \frac{\quad}{3}$

# Independent Work

## Part 4

**Copy the table. Figure out the missing numbers and answer the questions.**

This table is supposed to show the number of cats in two towns.

| | Black cats | Other cats | Total cats |
|---|---|---|---|
| West Town | 164 | 37 | |
| East Town | | | |
| Total for both towns | 210 | 270 | |

***Questions***

a. Are there more black cats in West Town or in East Town?

b. How many cats that are not black are in East Town?

c. Which town has more cats that are not black?

d. How many black cats are in both towns?

## Part 5

**Figure out the answer for each problem. Use your calculator to check the answers.**

a. $56 + 905 + 32 = ■$

b. $486 + 941 = ■$

c. $503 - 493 = ■$

d. $685 - 496 = ■$

## Part 6

**Figure out the number of square feet in each rectangle.**

a.

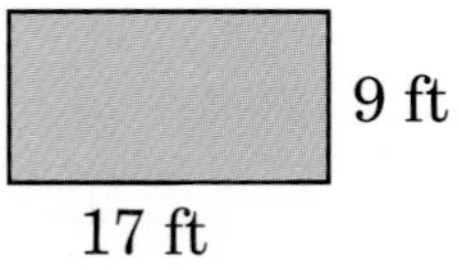


b.

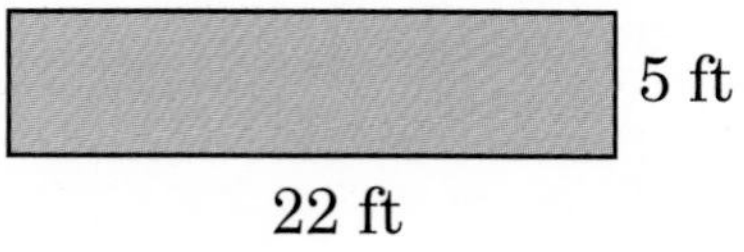

**Part 7** **Copy each problem and work it. Use a calculator to check the answers.**

a. $\begin{array}{r} 586 \\ \underline{\times \;\; 5} \end{array}$ b. $\begin{array}{r} 418 \\ \underline{\times \;\; 9} \end{array}$ c. $\begin{array}{r} 306 \\ \underline{\times \;\; 4} \end{array}$ d. $\begin{array}{r} 580 \\ \underline{\times \;\; 2} \end{array}$

**Part 8** **Copy each problem you can work and write the answer.**

a. $\frac{27}{3} - \frac{2}{3} =$ b. $\frac{5}{23} + \frac{5}{5} =$ c. $\frac{37}{12} + \frac{37}{1} =$

d. $\frac{70}{5} + \frac{20}{5} =$ e. $\frac{65}{64} - \frac{64}{64} =$ f. $\frac{56}{56} - \frac{1}{2} =$

**Part 9** **Write the division problem and the answer for each problem.**

a. 3 x ■ = 18 b. 9 x ■ = 63 c. 5 x ■ = 15 d. 3 x ■ = 27

**Part 10** **Copy the fractions that equal whole numbers. Write the complete equation to show the whole number each fraction equals.**

a. $\frac{7}{5}$ b. $\frac{7}{7}$ c. $\frac{7}{1}$ d. $\frac{12}{9}$ e. $\frac{16}{3}$ f. $\frac{27}{9}$

**Part J**

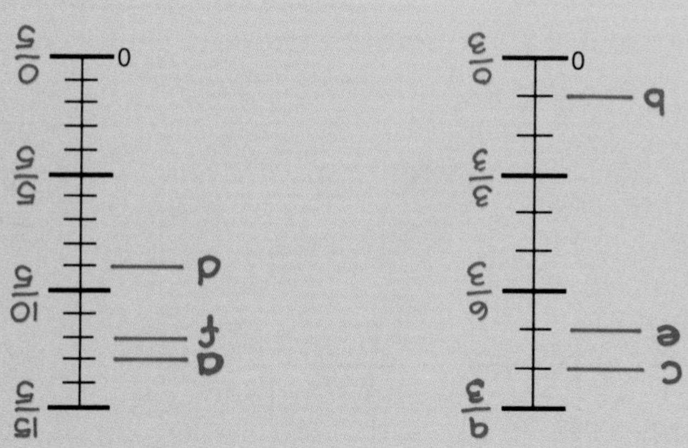

# Lesson 37

## Part 1

a. 14 = 9 + 5  b. 8 = 11 – 3  c. 18 = 9 + 9  d. 20 = 5 x 4

## Part 2

a. There were some crows and 278 other birds in a field. There were 1109 birds in all. How many crows were there?

b. Fran collected 340 more seashells than Andrea collected. Andrea has 555 seashells. How many does Fran have?

c. A horse drank 37 fewer cups of water than the camel drank. A camel drank 145 cups of water. How many cups of water did the horse drink?

d. 31 of the shirts on the clotheslines were wet, and the rest of the shirts were not wet. 299 shirts were not wet. How many shirts were on the clotheslines?

## Part 3

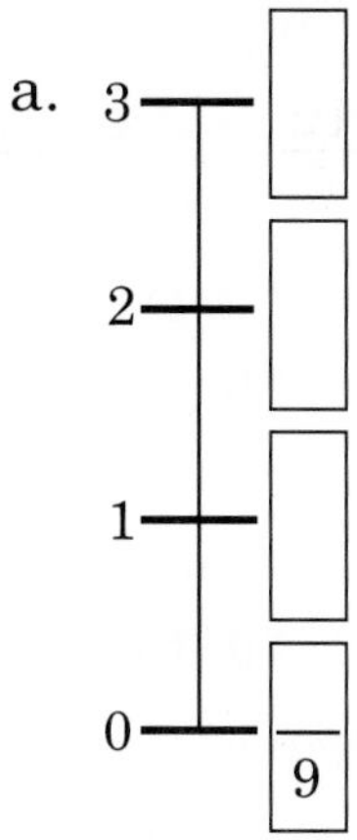


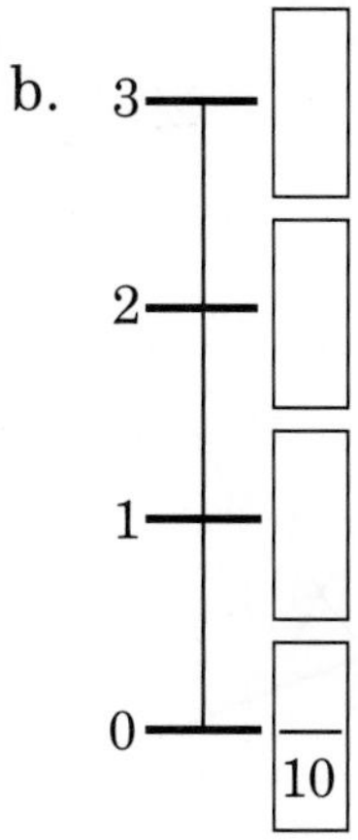


## Part 4

| | | | | |
|---|---|---|---|---|
| 4 | 8 | 12 | 16 | 20 |
| 24 | 28 | 32 | 36 | 40 |

| | | | | |
|---|---|---|---|---|
| 1 | 2 | 3 | 4 | 5 |
| 6 | 7 | 8 | 9 | 10 |

## Independent Work

### Part 5 Copy each problem and work it.

a. $\begin{array}{r} 74 \\ \underline{\times\ 50} \end{array}$ b. $\begin{array}{r} 48 \\ \underline{\times\ 30} \end{array}$ c. $\begin{array}{r} 48 \\ \underline{\times\ \ 3} \end{array}$ d. $\begin{array}{r} 8 \\ \underline{\times\ 30} \end{array}$

e. $\begin{array}{r} 9 \\ \underline{\times\ 20} \end{array}$ f. $\begin{array}{r} 92 \\ \underline{\times\ \ 6} \end{array}$ g. $\begin{array}{r} 4 \\ \underline{\times\ 50} \end{array}$ h. $\begin{array}{r} 361 \\ \underline{\times\ \ \ 2} \end{array}$

### Part 6 Write the numeral for each description.

a. 1 hundred thousand 74
b. 28 thousand 7 hundred 6
c. 8 thousand 9 hundred 40
d. 3 hundred 6 thousand 8
e. 5 hundred 20

### Part 7 Copy the table. Figure out the missing numbers and answer the questions.

This table is supposed to show the number of students that attend two schools.

| | Boys | Girls | Total students |
|---|---|---|---|
| Edison School | 192 | | 379 |
| Franklin School | | 163 | 353 |
| Total for both schools | | 350 | |

***Questions***

a. How many boys go to both schools?

b. Are there fewer boys or girls that attend Edison School?

c. Do more students attend Franklin School or Edison School?

## Part 8

**Write the fractions for items a through d. Write complete equations for e through j.**

a. The numbers for this fraction are 31 and 16. The fraction is less than 1.

b. The numbers for this fraction are 6 and 16. The fraction is more than 1.

c. The picture shows 8 parts shaded. There are 5 parts in each unit.

d. 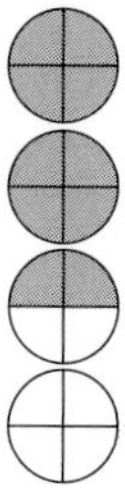

e. $7 = \frac{\quad}{3} = \frac{\quad}{5}$

f. $\frac{15}{3} + \frac{8}{3} =$

g. $\frac{15}{3} - \frac{8}{3} =$

h. $\frac{15}{3} =$

i. $9 = \frac{\quad}{1}$

j. $\frac{8}{8} =$

## Part 9

**For each item, write the addition or multiplication fact.**

a. 14, 1, 13 b. 10, 20, 2 c. 1, 1, 0 d. 8, 9, 72

## Part J

a. $2\overline{)640}$ 320

b. $3\overline{)276}$ 92

c. $5\overline{)205}$ 41

d. $2\overline{)164}$ 82

e. $9\overline{)909}$ 101

f. $3\overline{)189}$ 63

## Part K

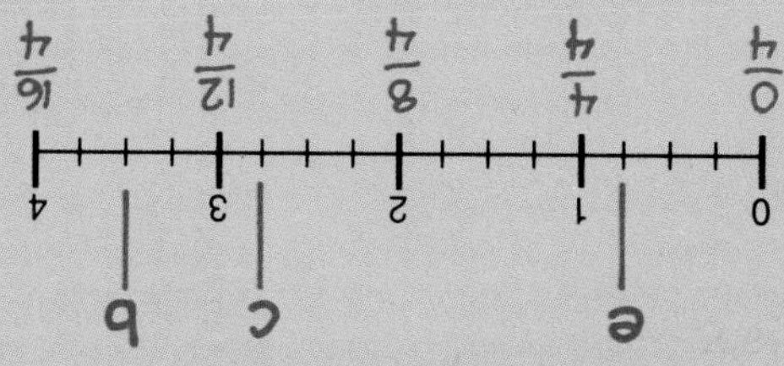

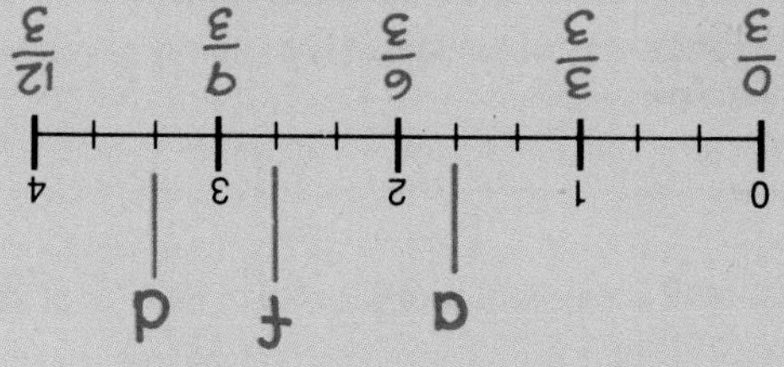

# Lesson 38

## Part 1

a. $3\overline{)24}$  b. $3\overline{)27}$  c. $3\overline{)6}$  d. $3\overline{)15}$  e. $3\overline{)21}$

f. $3\overline{)3}$  g. $3\overline{)12}$  h. $3\overline{)18}$  i. $3\overline{)9}$

## Part 2

a. 15 = 9 + 6  b. 7 = 9 – 2  c. 24 = 4 x 6  d. 36 = 4 x 9

## Part 3

a.

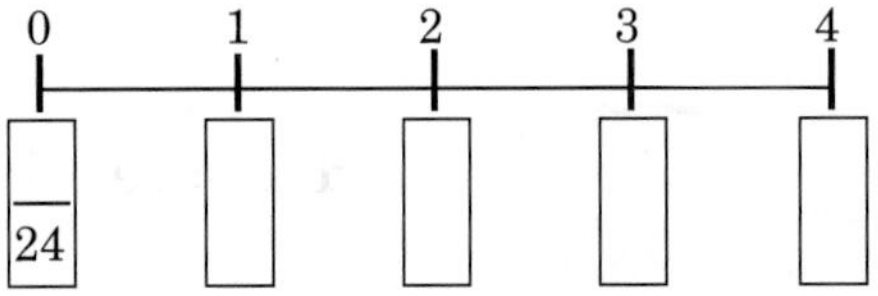


b.

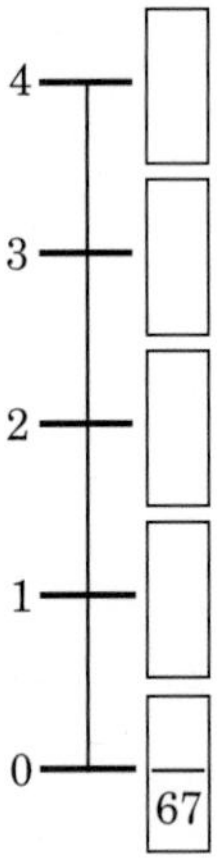


## Independent Work

## Part 4

**Write the division problem and the answer for each problem.**

a. 9 x ■ = 81  b. 2 x ■ = 6  c. 3 x ■ = 18  d. 5 x ■ = 40

## Part 5

**Copy each problem you can work and write the answer.**

a. $\frac{4}{17} + \frac{20}{17} =$

b. $\frac{4}{6} + \frac{4}{20} =$

c. $\frac{40}{9} + \frac{30}{8} =$

d. $\frac{7}{9} + \frac{30}{9} =$

e. $\frac{30}{2} - \frac{30}{2} =$

f. $\frac{20}{10} + \frac{20}{20} =$

## Part 6

**For each problem, make the number family and figure out the answer. Some families have a difference number.**

a. Snow Mountain is 381 feet higher than Rock Mountain. Rock Mountain is 9006 feet high. How many feet high is Snow Mountain?

b. Jill had some pennies and 130 coins that were not pennies. Jill had 484 coins in all. How many pennies did Jill have?

c. A book has 170 more pages than a catalog. The book has 643 pages. How many pages does the catalog have?

d. Reggie caught 89 balls and Derek caught 62 balls. How many total balls did Reggie and Derek catch?

## Part 7

**For each problem, draw a diagram and answer the question.**

a. A wall is 8 feet high and 23 feet wide. What is the area of the wall?

b. A piece of paper is 17 inches long and 5 inches wide. What is the area of the paper?

# Lesson 39

## Part 1

This table shows the number of parrots and monkeys in Rover Park.

| | Parrots | Monkeys | Total for both animals |
|---|---|---|---|
| Rover Park | 50 | 70 | 120 |

***Questions***

a. In Rover Park, there are 45 more swallows than monkeys. How many swallows are in Rover Park?

b. In Rover Park, there are 26 fewer bears than parrots. How many bears are in Rover Park?

c. In Rover Park, the total number of monkeys and parrots is 56 less than the number of robins. How many robins are in Rover Park?

## Part 2

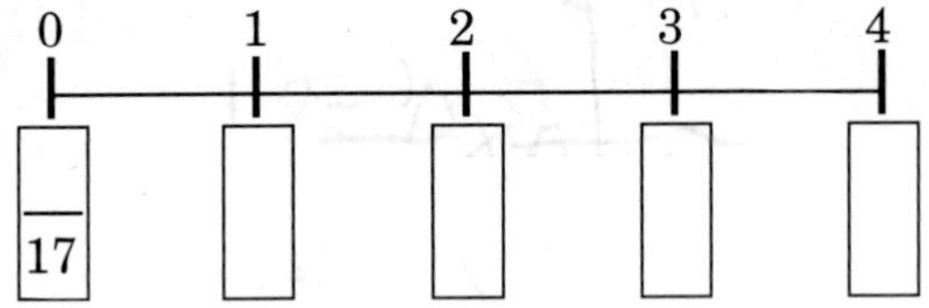


## Part 3

a. $3 + \frac{2}{5}$ b. $7 + \frac{1}{9}$ c. $6 + \frac{3}{4}$

## Independent Work

### Part 4

**Write the addition or multiplication fact for each group of numbers.**

a. 8, 5, 3    b. 4, 28, 7    c. 3, 15, 5    d. 11, 6, 5

### Part 5

**Write the time shown on each clock.**

a.

b.

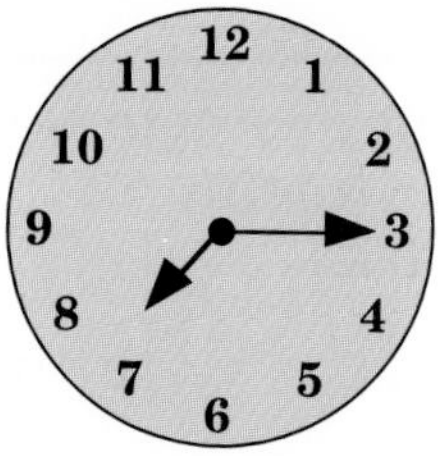


c.

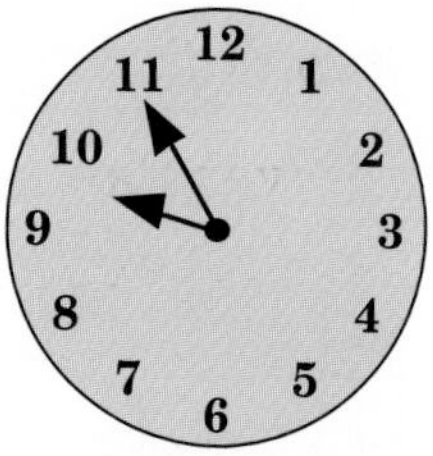


### Part 6

**Copy the table. Figure out the missing numbers. Answer the questions.**

This table is supposed to show the number of good bricks and chipped bricks at two different companies.

| | Good bricks | Chipped bricks | Total bricks |
|---|---|---|---|
| Ace Builders | 72 | | 397 |
| Sam's Construction | | 330 | 736 |
| Total for both companies | 478 | | |

***Questions***

a. Which company had more good bricks?

b. How many bricks did Sam's Construction have in all?

c. How many chipped bricks were there?

d. Were there fewer chipped bricks or good bricks?

## Part 7 Make a number family and answer the question.

a. Jim has money in his wallet and the rest of his money in a jar. He has $14.75 in the jar. If Jim has $65.80 in all, how much money does he have in his wallet?

## Part 8 Copy each problem and work it.

| a. | b. | c. | d. | e. |
|---|---|---|---|---|
| 12 | 76 | 8 | 78 | 43 |
| x 90 | x 3 | x 30 | x 4 | x 60 |

## Part 9 Write the addition problem for each item and figure out how much each person spent.

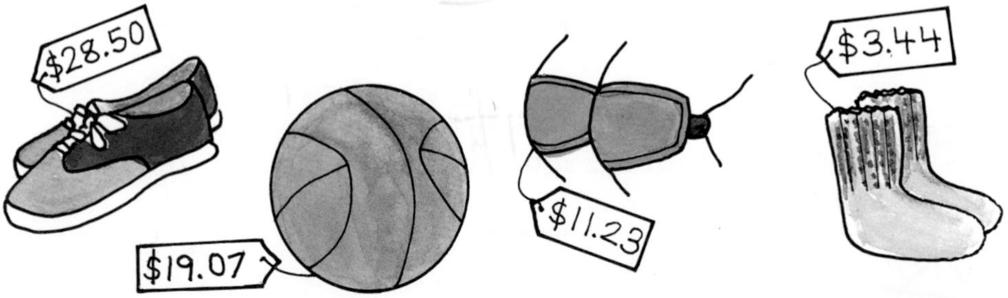

$28.50
$19.07
$11.23
$3.44

a. Jill bought the basketball and the kneepads. How much did she spend?

b. Steve bought the shoes, the kneepads and the socks. How much did he spend?

$14.85
+6.50
Alex, that's not how you line up dollar amounts.

# Test 4

## Part 8

a. 67 of the birds were hungry, and the rest were not hungry. There were 341 birds in all. How many were not hungry?

b. The kite was 34 feet higher than a bird. The kite was 78 feet high. How high was the bird?

## Part 9

a. $\frac{7}{3} + \frac{3}{3} =$ b. $\frac{13}{6} - \frac{8}{6} =$ c. $\frac{11}{7} - \frac{11}{9} =$

d. $\frac{22}{10} + \frac{3}{10} =$ e. $\frac{1}{2} - \frac{1}{2} =$ f. $\frac{7}{8} + \frac{7}{7} =$

# Lesson 40

## Part 1

a. $2 + \frac{8}{10}$ b. $7 + \frac{2}{9}$ c. $8 + \frac{1}{4}$ d. $3 + \frac{0}{5}$

## Part 2

a. $\frac{\quad}{3} = 6$ b. $\frac{\quad}{4} = 8$ c. $\frac{\quad}{2} = 4$ d. $\frac{\quad}{7} = 10$

# Lesson 41

## Part 1

a. $\frac{\quad}{9} = 5$ b. $\frac{\quad}{3} = 7$ c. $\frac{\quad}{1} = 8$ d. $\frac{\quad}{6} = 2$

## Part 2

This table shows the number of cars and trucks on Ted's lot.

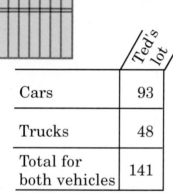

| | Ted's lot |
|---|---|
| Cars | 93 |
| Trucks | 48 |
| Total for both vehicles | 141 |

***Questions***

a. There are 71 fewer cars on the street than cars on Ted's lot. How many cars are on the street?

b. On Ted's lot, there are 65 more bicycles than trucks. How many bicycles are on Ted's lot?

c. There are 17 more vehicles on Ted's lot than there are on Sally's lot. How many vehicles are on Sally's lot?

## Part 3

a. $7 + \frac{1}{2}$ b. $2 + \frac{7}{9}$

c. $5 + \frac{5}{8}$ d. $3 + \frac{2}{4}$

## Part 4

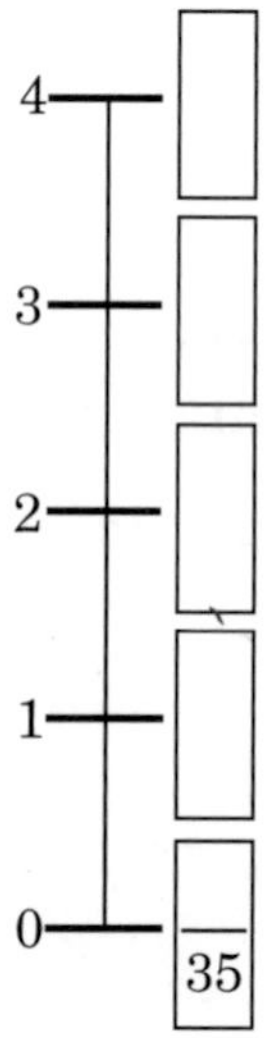

## Part 5

- Some problems multiply by 2-digit values that do not end in zero.
- This problem multiplies by 24:

```
  3 2
x 2 4
```

- To get the answer, you have to work two problems. If you cover the first digit of 24, you can see the first problem you work.
- You multiply 32 x 4. The answer is 128. You write the answer right below the problem.

```
  3 2
x ■ 4
1 2 8
```

- If you cover the second digit of 24, you can see the second problem you work: 32 x 20. There's a zero in the ones column of the answer.
- You write the answer below the answer for 32 x 4. The answer is 640. You write the answer here: ⟶

```
  3 2
x 2 ■
1 2 8
6 4 0
```

- To find the answer to the whole problem, you add 128 and 640. The answer to the whole problem is 768.

```
   3 2
x  2 4
 1 2 8
+6 4 0
 7 6 8
```

## Independent Work

## Part 6

**Make a number family for each problem. Figure out the missing number and write the answer.**

a. 46 people in a school were teachers, and the rest of the people were students. There were 196 students in the school. How many total people were in the school?

b. 97 of the animals in a field were cows. The rest were birds. There were 116 animals in the field. How many birds were in the field?

## Part 7

**Copy each problem and work it.**

a. $\begin{array}{r} 18 \\ \underline{\times\ 20} \end{array}$  b. $\begin{array}{r} 978 \\ \underline{\times\ \ \ 4} \end{array}$

## Part 8

**Rewrite each equation so the number after the equal sign is first.**

a. 8 + 6 = 14  b. 1 + 9 = 10  c. 4 x 9 = 36  d. 5 x 3 = 15

## Part 9

**Write the division problem and the answer for each problem.**

a. 3 x ■ = 18  b. 5 x ■ = 35  c. 3 x ■ = 12  d. 9 x ■ = 27

e. 3 x ■ = 21  f. 5 x ■ = 20  g. 2 x ■ = 14  h. 3 x ■ = 24

## Part 10

**Copy each problem you can work and write the answer.**

a. $\frac{30}{5} - \frac{29}{5} =$  b. $\frac{50}{9} + \frac{30}{90} =$  c. $\frac{27}{15} - \frac{3}{15} =$

d. $\frac{2}{3} + \frac{27}{2} =$  e. $\frac{20}{15} + \frac{70}{15} =$  f. $\frac{4}{9} - \frac{3}{9} =$

## Part 11

**A sign or a number is missing in each problem. Copy each problem and complete the equation. The equal signs for some equations come after the first number.**

a. 16 = 10 ■ 6  b. 10 = 5 x ■  c. 15 + ■ = 16  d. 35 = 40 ■ 5

# Lesson 42

## Part 1

a. $2 + \frac{5}{9}$ b. $\frac{3}{8} + \frac{12}{8}$ c. $\frac{15}{4} - \frac{6}{4}$ d. $4 + \frac{7}{10}$

## Part 2

| | | | | |
|---|---|---|---|---|
| 1 | 2 | 3 | 4 | 5 |
| 6 | 7 | 8 | 9 | 10 |

a. $4\overline{)28}$ b. $4\overline{)8}$ c. $4\overline{)36}$ d. $4\overline{)24}$

e. $4\overline{)12}$ f. $4\overline{)20}$ g. $4\overline{)32}$

## Part 3

This table shows the number of green cones and ripe cones on a pine tree.

| | Pine tree |
|---|---|
| Green cones | 56 |
| Ripe cones | 160 |
| Total cones | 216 |

***Questions***

a. The number of ripe cones is 34 less than the number of ants on the pine tree. How many ants are on the tree?

b. A cedar tree has 530 more green cones than the pine tree has. How many green cones are on the cedar tree?

c. The number of cones on the ground is 43 less than the total cones on the pine tree. How many cones are on the ground?

## Part 4

- If **fractions** are **equal,** pictures of the fractions **have the same shaded area.**
- Here are two fractions that are equal: $\frac{2}{3} = \frac{8}{12}$
- Here are the pictures:

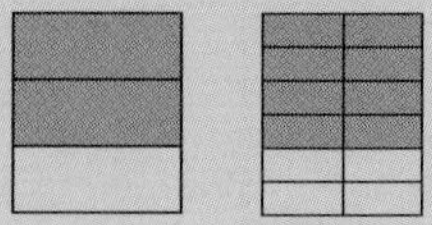

- Both pictures have a shaded area that is the same size and an unshaded area that is the same size. The pictures show that the fractions are equal.
- If the **fractions** are **not equal,** the pictures **do not have the same shaded area.**
- Here are two fractions that are not equal: $\frac{2}{3}$ $\frac{10}{12}$
- $\frac{2}{3}$ does not equal $\frac{10}{12}$. The picture of $\frac{10}{12}$ has a larger shaded area and a smaller area that is not shaded.

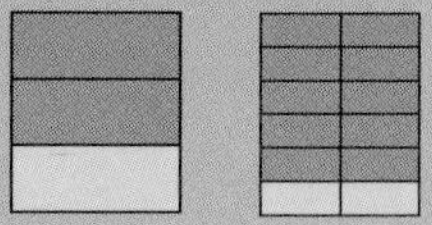

## Independent Work

## Part 5

**Write the column problem for each item and work it. Use your calculator to check the answers.**

a. 406 x 3 = ■ b. 941 x 60 = ■ c. 121 x 9 = ■

## Part 6

**Write the division problem and the answer for each problem.**

a. 9 x ■ = 36 b. 5 x ■ = 35 c. 6 x ■ = 6 d. 2 x ■ = 8

**Part 7** **Copy the table. Figure out all the missing numbers.**

| | | | Total |
|---|---|---|---|
| | 132 | | 301 |
| | | 41 | |
| Total | 732 | 210 | |

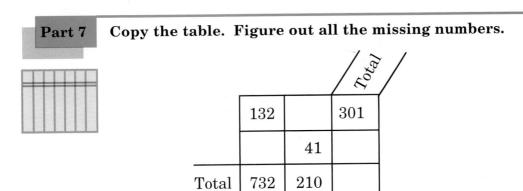

Are you sure you were supposed to make that kind of table?
WOOD GLUE

# Lesson 43

## Part 1

- You've worked number-line problems with fractions that have 2-digit values.
- The fraction for the whole number 3 has a top number that is 3 times the bottom number.

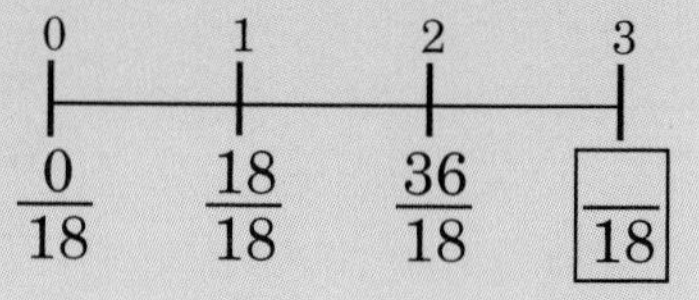


- If the bottom number is 18, you find the top number by multiplying 3 x 18. The top number is 54.

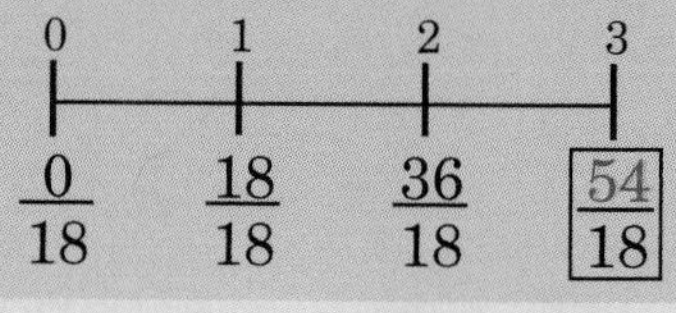


- Here's the same problem without the number line:

$$3 = \frac{\quad}{18}$$

- The problem tells you that 3 equals a fraction that has a bottom number of 18. If the fraction equals 3, the top number is 3 times the bottom number. That's 54.

$$3 = \frac{54}{18}$$

a. $4 = \frac{\quad}{12}$

b. $9 = \frac{\quad}{12}$

c. $6 = \frac{\quad}{50}$

## Part 2

- These are problems that **ask** about the **difference.**
- Here's a problem: **A coat cost $134. A radio cost $23. What is the difference in the price of these objects?**
- The question asks about the difference, so you start with a number family that compares. You can figure out which is the big number by comparing the cost of the **coat** and the **radio.** The coat costs more. It's the big number.
- Here's the number family:

| difference | radio | coat |
|---|---|---|
| | 23 | → 134 |

- To find the difference number, you subtract: 134 – 23. That's 111. The difference in price is $111.

$$\begin{array}{r} \$134 \\ -\ \ 23 \\ \hline \$111 \end{array}$$

a. Joan is 56 inches tall. Ted is 37 inches tall. What is the difference in their height?

b. The bike went 16 miles per hour. The car went 75 miles per hour. What was the difference in the speed of the vehicles?

c. A tub held 112 gallons. A tank held 256 gallons. What was the difference in the amounts held by the containers?

## Part 3

a. $7 - \frac{2}{3}$  b. $\frac{12}{5} - 2$  c. $\frac{17}{3} - 3$  d. $\frac{3}{5} + 7$

## Part 4

a. $\begin{array}{r} 43 \\ \underline{\times\ 23} \end{array}$  b. $\begin{array}{r} 82 \\ \underline{\times\ 54} \end{array}$  c. $\begin{array}{r} 74 \\ \underline{\times\ 92} \end{array}$

## Part 5

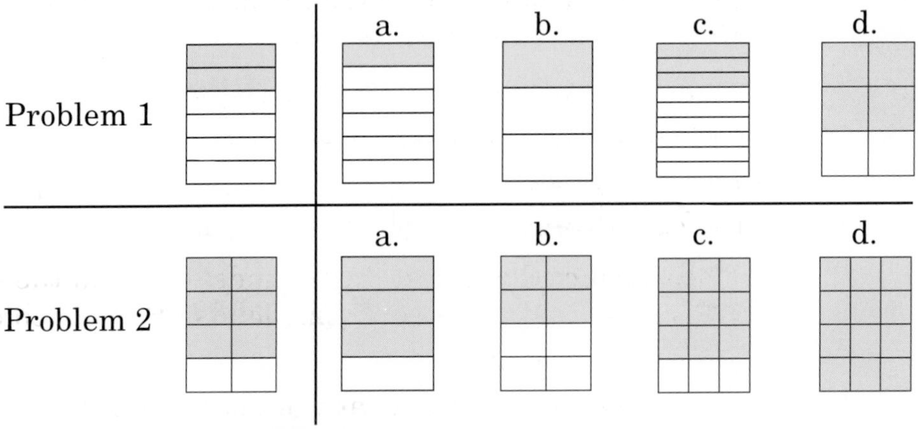


## Independent Work

## Part 6

**For each problem, make a diagram and answer the question.**

a. A strip of land is 9 feet wide and 38 feet long. What is the area of the strip?

b. A poster is 50 inches high and 16 inches wide. What is the area of the poster?

## Part 7

**Copy each problem and work it. Underline the digits for the first digit of the answer.**

a. $3\overline{)276}$ b. $4\overline{)804}$ c. $2\overline{)804}$ d. $3\overline{)120}$

**Part 8** **Make a number family for each problem. Figure out the missing number and write the answer.**

a. In Joe's Diner, there were 34 dirty plates and 152 clean plates. How many plates were there in all?

b. A pine tree is 45 feet taller than a maple tree. The pine tree is 171 feet tall. How tall is the maple tree?

c. The pile of coal weighed 251 pounds less than the sand pile. The pile of coal weighed 1222 pounds. How much did the pile of sand weigh?

d. 23 of the people on the boat wore hats, and the rest did not wear hats. There was a total of 88 people on the boat. How many did not wear hats?

**Part 9** **Rewrite each equation so the number after the equal sign is first.**

a. 8 + 6 = 14　　b. 3 x 7 = 21　　c. 2 x 9 = 18　　d. 14 – 9 = 5

# Lesson 44

## Part 1

a. $\frac{15}{4} - \frac{7}{4}$

b. $\frac{2}{9} + 5$

c. $2 + \frac{4}{7}$

d. $\frac{26}{2} - 3$

e. $\frac{3}{5} + 8$

## Part 2

a. Jan was 67 inches tall. Rita was 46 inches tall. What was the difference in the height of the two girls?

b. The car traveled 134 miles. The bus traveled 340 miles. How much farther did the bus travel than the car traveled?

c. Train car A weighed 76 tons. Train car B weighed 39 tons. How much lighter was car B than car A?

## Part 3

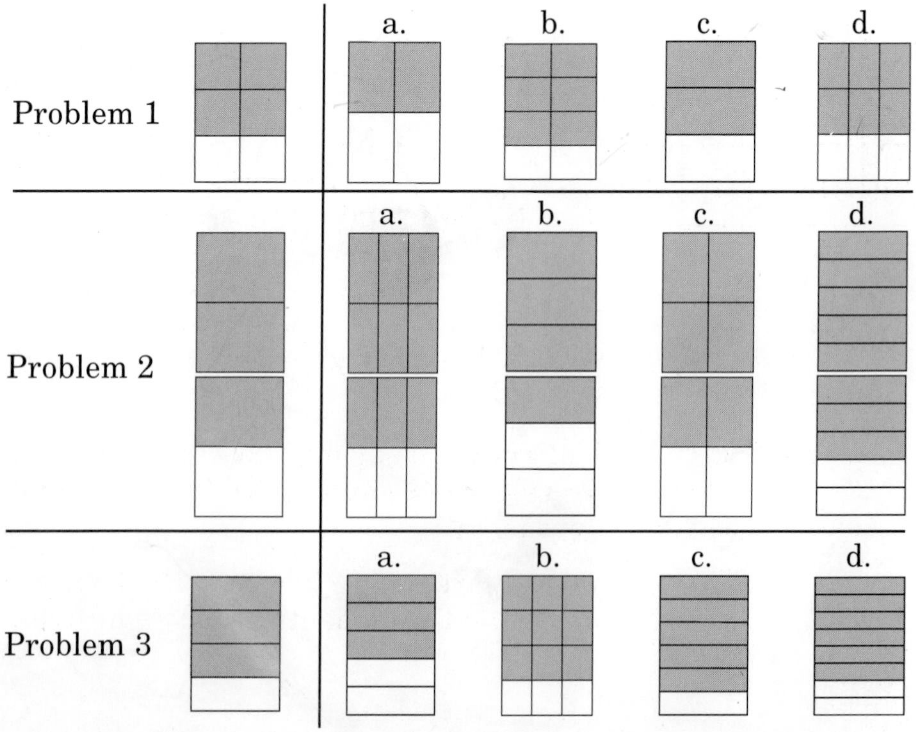

## Independent Work

**Part 4** **Copy the table. Figure out all the missing numbers.**

| | | | Total |
|---|---|---|---|
| | 11 | | |
| | | 149 | 249 |
| Total | 111 | | 360 |

**Part 5** **Use the information to figure out the answers to the problems.**

This table shows the number of frogs and fish that are in a pond.

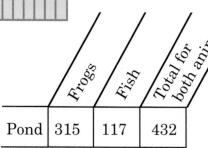

| | Frogs | Fish | Total for both animals |
|---|---|---|---|
| Pond | 315 | 117 | 432 |

***Questions***

a. There are 134 fewer beetles than frogs. How many beetles are in the pond?

b. The number of water spiders is 33 more than the number of fish in the pond. How many water spiders are there?

**Part 6** **Make a number family for each problem. Figure out the missing number and write the answer.**

a. 12 acres of a field were burnt. The rest of the field was not burnt. If there were 289 acres in the whole field, how many acres were not burnt?

b. There were 1113 cars on Dill Street. 950 of the cars were not dented. The rest were dented. How many cars were dented?

## Part 7

**Copy each problem and work it. Use your calculator to check the answers.**

a. $\begin{array}{r} 470 \\ \times\ \ 20 \\ \hline \end{array}$

b. $\begin{array}{r} 378 \\ -\ 129 \\ \hline \end{array}$

c. $\begin{array}{r} 564 \\ -\ 188 \\ \hline \end{array}$

d. $\begin{array}{r} 960 \\ \times\ \ \ 2 \\ \hline \end{array}$

e. $\begin{array}{r} 564 \\ +\ 188 \\ \hline \end{array}$

f. $\begin{array}{r} 39 \\ 406 \\ +\ \ 12 \\ \hline \end{array}$

## Part 8

**Copy each problem you can work and write the answer.**

a. $\frac{7}{3} - \frac{7}{5} =$

b. $\frac{2}{11} + \frac{15}{11} =$

c. $\frac{6}{4} - \frac{5}{5} =$

d. $\frac{20}{7} - \frac{7}{7} =$

e. $\frac{8}{9} + \frac{9}{8} =$

f. $\frac{6}{2} - \frac{6}{5} =$

## Part 9

**Rewrite each equation so the number after the equal sign is first.**

a. $100 - 17 = 83$

b. $64 \times 4 = 256$

## Part 10

**Copy each problem and complete the equation.**

a. $3 = \frac{\quad}{57}$

b. $2 = \frac{\quad}{1}$

c. $9 = \frac{\quad}{16}$

d. $\frac{\quad}{4} = 20$

# Lesson 45

**Part 1**

a. $\frac{13}{4} - \frac{13}{7}$

b. $\frac{26}{3} - 7$

c. $\frac{10}{4} - \frac{4}{3}$

d. $\frac{10}{4} + \frac{36}{4}$

e. $8 + \frac{6}{5}$

f. $\frac{13}{2} - \frac{3}{4}$

g. $\frac{16}{2} + 5$

h. $\frac{1}{2} - \frac{1}{6}$

i. $\frac{5}{5} - \frac{5}{8}$

**Part 2**

a. Green Road is 13 miles shorter than Beach Road. If Green Road is 85 miles long, how long is Beach Road?

b. Blake weighs 32 pounds. Her mother weighs 124 pounds. How much lighter is Blake than her mother?

c. A maple tree is 88 feet tall. An oak tree is 39 feet tall. How much taller is the maple than the oak?

d. An elephant was 112 years younger than a turtle. The turtle was 134 years old. How old was the elephant?

**Part 3**

- When you divide, you write the first digit of the answer. Then you must write a digit in the answer above every digit under the division sign.
- Here's a problem:
  The first digit of the answer is above the 2. There's a digit in the answer above each of the digits under the division sign.

  **400**
  **3 | 1200**

- Here's the wrong way:
  There's no digit in the answer above the last zero.

  **40**
  **3 | 1200**

a. 9 | 6309 (answer: 701)

b. 3 | 900 (answer: 3 _ 0)

c. 5 | 4505 (answer: 90)

d. 5 | 3505 (answer: 701)

e. 9 | 2700 (answer: 3 _ 0)

f. 3 | 6093 (answer: 2031)

g. 4 | 200 (answer: 5)

# Independent Work

## Part 4

**Write the addition problem and answer each question.**

a. Bob buys the hat and sunglasses. How much does he spend?

b. Jill buys shoes, the hat and gloves. How much does she spend?

c. Amy buys shoes and sunglasses. How much does she spend?

## Part 5

**Copy each problem and work it.**

a. $\begin{array}{r} 463 \\ \underline{x\ \ 20} \end{array}$ b. $\begin{array}{r} 463 \\ \underline{x\ \ 90} \end{array}$ c. $\begin{array}{r} 518 \\ \underline{x\ \ \ \ 5} \end{array}$ d. $\begin{array}{r} 409 \\ \underline{x\ \ 60} \end{array}$ e. $\begin{array}{r} 210 \\ \underline{x\ \ \ \ 8} \end{array}$

## Part 6

**Copy each problem and complete the equation.**

a. $21 = \frac{\quad}{3}$ b. $\frac{\quad}{12} = 4$ c. $14 = \frac{\quad}{5}$

## Part 7

**Write the time shown on each clock.**

a.

b.

**Part 8** **Write the fraction for each description. Circle all fractions that are more than 1.**

a. The picture shows 20 parts in each unit. 23 are shaded.

b. 11 parts are shaded. There are 18 parts in each unit.

c. The fraction is more than 1. The numbers are 300 and 400.

d. The fraction is less than 1. The numbers are 31 and 13.

**Part 9** **Write the fraction for each picture.**

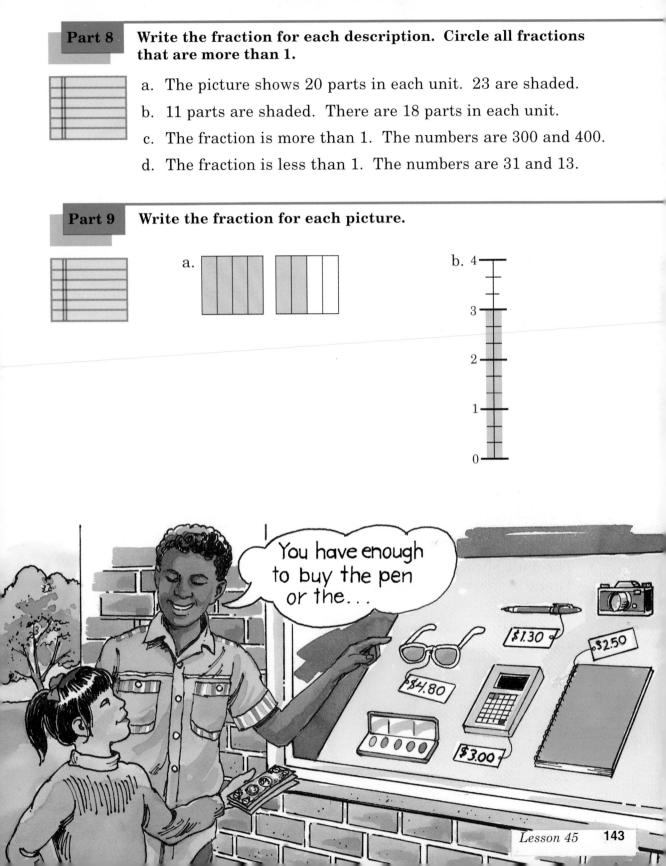

a.

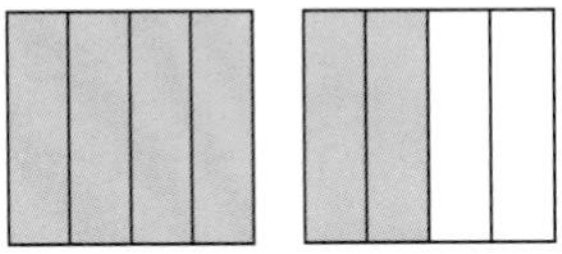

b.

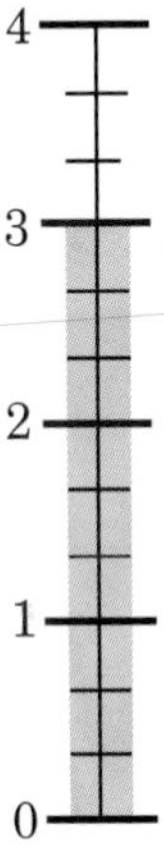

# Lesson 46

## Part 1

a. Tim had $35.80. He bought the helmet. How much did he end up with?

b. Tina had $241.00. She bought the bike. How much did she end up with?

c. Reggie had $20.75. He bought the backpack. How much did he end up with?

## Part 2

- When you multiply fractions, you multiply the top numbers and write the answer on top. Then you multiply the bottom numbers and write the answer on the bottom.

Here's: $\frac{3}{4} \times \frac{2}{5}$

- The multiplication problem for the top is 3 x 2. That's 6. You write 6 on top. $\frac{3}{4} \times \frac{2}{5} = \frac{6}{\square}$

- The multiplication problem for the bottom is 4 x 5. That's 20. You write 20 on the bottom. $\frac{3}{4} \times \frac{2}{5} = \frac{6}{20}$

a. $\frac{2}{9} \times \frac{4}{2} = \square$

b. $\frac{2}{3} \times \frac{4}{1} = \square$

c. $\frac{7}{2} \times \frac{2}{10} = \square$

d. $\frac{1}{3} \times \frac{5}{2} = \square$

e. $\frac{2}{8} \times \frac{4}{1} = \square$

## Part 3

a. An oak tree is 345 years old. A pine tree is 178 years old. How much younger is the pine tree than the oak tree?

b. An oak tree is 167 years old. Mr. Brown is 134 years younger than the oak tree. How old is Mr. Brown?

c. Fran ran 7891 yards. Jill ran 333 yards farther than Fran. How far did Jill run?

d. Jill was 3601 days old. Fran was 4012 days old. How many days older than Jill was Fran?

# Independent Work

## Part 4

**Copy each problem and write the answer.**

a. $4\overline{)24}$ b. $4\overline{)16}$ c. $3\overline{)15}$ d. $3\overline{)24}$

e. $4\overline{)32}$ f. $4\overline{)36}$ g. $3\overline{)21}$ h. $4\overline{)28}$

## Part 5

**Copy each problem and work it.**

a. $3 - \frac{6}{9}$ b. $\frac{10}{5} - \frac{10}{5}$ c. $\frac{11}{3} + \frac{2}{3}$ d. $\frac{8}{4} + 1$

## Part 6

**For each problem, make a diagram and answer the question.**

a. A field is 15 yards wide and 9 yards long. What's the area of the field?

b. A wall is 8 feet tall and 43 feet wide. What is the area of the wall?

**Part 7** **Write the number for each description.**

a. 5 thousand 2 hundred 15

b. 71 thousand 1 hundred

c. 6 thousand 8 hundred 4

d. 12 thousand 6

**Part 8** **Copy each problem and work it. Underline the digits for the first digit of the answer.**

a. $3\overline{)1506}$

b. $5\overline{)5005}$

c. $4\overline{)8004}$

d. $4\overline{)160}$

**Part 9** **Copy the table. Figure out the missing numbers. Answer the questions.**

This table is supposed to show the number of flowering plants in Tintown and Vistaville.

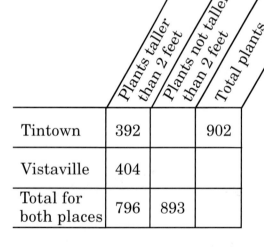

| | Plants taller than 2 feet | Plants not taller than 2 feet | Total plants |
|---|---|---|---|
| Tintown | 392 | | 902 |
| Vistaville | 404 | | |
| Total for both places | 796 | 893 | |

***Questions***

a. Which place had more plants taller than 2 feet?

b. Which place had the most plants?

c. What was the total number of plants that were not taller than 2 feet?

d. What was the total number of all plants?

# Lesson 47

## Part 1

a. $\frac{2}{3} \times \frac{7}{5} = \frac{\square}{\square}$   b. $\frac{5}{2} \times \frac{1}{6} = \frac{\square}{\square}$   c. $\frac{3}{5} \times \frac{2}{4} = \frac{\square}{\square}$

d. $\frac{5}{8} \times \frac{7}{2} = \frac{\square}{\square}$   e. $\frac{2}{7} \times \frac{7}{2} = \frac{\square}{\square}$

## Part 2

a. How much more does the hat cost than the broom costs?

b. How much more does the skateboard cost than the running shoes costs?

c. How much more does the skateboard cost than the hat costs?

d. How much less is the bucket than the hat?

## Part 3

a. $3 - \frac{1}{8} =$   b. $\frac{7}{18} - \frac{7}{16} =$   c. $\frac{12}{30} + \frac{2}{30} =$

d. $\frac{2}{5} + 9 =$   e. $\frac{12}{7} + \frac{12}{9} =$   f. $\frac{11}{2} + \frac{34}{2} =$

# Independent Work

## Part 4

**Copy each problem and work it. Underline the digits for the first digit of the answer.**

a. $9\overline{)540}$  b. $9\overline{)90}$  c. $9\overline{)900}$  d. $9\overline{)450}$

e. $4\overline{)8808}$  f. $4\overline{)200}$  g. $4\overline{)244}$

## Part 5

**Copy the table and complete it. Work the problems for each question.**

| | Older | Newer | Total |
|---|---|---|---|
| Cars on M Street | 56 | 14 | |

***Questions***

a. On M Street, there are 61 more bushes than there are cars. How many bushes are on M Street?

b. On L Street, there are 23 fewer cars than there are on M Street. How many cars are on L Street?

c. The number of newer cars on L Street is 3 less than the number of newer cars on M Street. How many newer cars are there on L Street?

## Part 6

**Copy each problem and work it.**

a. $56 \times 90$  b. $82 \times 5$  c. $37 \times 40$

## Part 7

**Copy each problem and complete the equation.**

a. $3 = \frac{\quad}{15}$  b. $135 = \frac{\quad}{5}$  c. $24 = \frac{\quad}{2}$  d. $8 = \frac{\quad}{94}$

# Lesson 48

## Part 1

a. $\begin{array}{r} 83 \\ \underline{\times\ 94} \end{array}$

b. $\begin{array}{r} 56 \\ \underline{\times\ 53} \end{array}$

c. $\begin{array}{r} 78 \\ \underline{\times\ 43} \end{array}$

## Part 2

a. $\frac{2}{4} \times \frac{3}{1} = \frac{\square}{\square}$

b. $\frac{4}{3} \times \frac{3}{5} = \frac{\square}{\square}$

c. $\frac{9}{5} \times \frac{3}{6} = \frac{\square}{\square}$

d. $\frac{2}{10} \times \frac{5}{1} = \frac{\square}{\square}$

e. $\frac{7}{5} \times \frac{2}{3} = \frac{\square}{\square}$

## Part 3

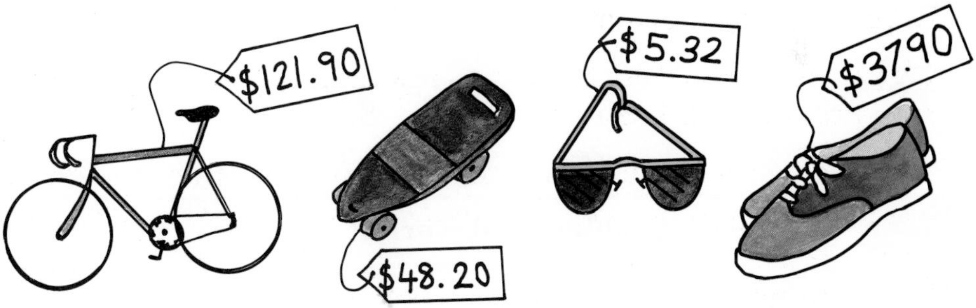


a. Tom wants to buy the skateboard. Tom has $34.20. How much more money does he need?

b. Fran has $25.00. She wants to buy running shoes. How much more money does she need?

c. Linda has $14.85. She wants to buy the bike. How much more money does she need?

d. Alex wants to buy sunglasses. He has $4.18. How much more money does he need?

**Part 4**

This number map shows the numbers for counting by 7. The second digit of each red number is 1 more than the digit above it.

| | | |
|---|---|---|
| 7 | 14 | 21 |
| 28 | 35 | 42 |
| 49 | 56 | 63 |
| 70 | | |

## Independent Work

**Part 5** **Copy each item and complete the equation.**

a. $4 = \frac{}{7} = \frac{}{5} = \frac{}{9} = \frac{}{3}$

b. $2 = \frac{}{1} = \frac{}{6} = \frac{}{2} = \frac{}{4}$

c. $1 = \frac{}{1} = \frac{}{8} = \frac{}{10} = \frac{}{47}$

**Part 6** **Copy each problem and work it.**

a. $\frac{5}{3} - \frac{2}{3} =$

b. $\frac{5}{3} + \frac{2}{3} =$

c. $\frac{5}{3} + 2 =$

d. $\frac{1}{6} + \frac{3}{6} =$

e. $\frac{13}{4} + \frac{83}{4} =$

f. $\frac{18}{3} - 4 =$

**Part 7** **Copy each problem and work it. Underline the digits for the first digit of the answer.**

a. $4\overline{)128}$

b. $4\overline{)804}$

c. $2\overline{)126}$

d. $3\overline{)609}$

## Part 8 Write the division problem and the answer for each problem.

a. 7 x ■ = 35   b. 8 x ■ = 72   c. 4 x ■ = 28

d. 5 x ■ = 35   e. 9 x ■ = 54

## Part 9 For each problem, make a number family and answer the questions.

a. The elephant was 17 years younger than the camel. The camel was 66 years old. How old was the elephant?

b. A grain shed held wheat and corn. There were 1084 pounds of corn in the shed. There were 1968 pounds grain in all. How many pounds of wheat were in the shed?

c. Tim sold 896 tickets. Donna sold 548 tickets. How many fewer tickets did Donna sell than Tim?

d. There were 87 empty bottles, and the rest were full bottles. If 450 bottles were full, how many bottles were there in all?

## Part 10 Copy each problem and work it. Use your calculator to check the answers.

a. $\begin{array}{r} 365 \\ \underline{\times\ 20} \end{array}$   b. $\begin{array}{r} 228 \\ \underline{-\ 199} \end{array}$   c. $\begin{array}{r} 356 \\ \underline{-\ \ 90} \end{array}$

d. $\begin{array}{r} 471 \\ \underline{+\ 1829} \end{array}$   e. $\begin{array}{r} 538 \\ \underline{+\ 394} \end{array}$   f. $\begin{array}{r} 376 \\ \underline{\times\ 30} \end{array}$

# Lesson 49

## Part 1

a. 186 × 49

b. 253 × 92

c. 768 × 35

## Part 2

a. $\frac{3}{1} \times \frac{1}{3} = \frac{\square}{\square}$

b. $\frac{7}{4} \times \frac{2}{3} = \frac{\square}{\square}$

c. $\frac{15}{8} \times \frac{1}{2} = \frac{\square}{\square}$

## Part 3

- Here's a new kind of problem: $3\overline{)1818}$
- You start by figuring out where to write the first digit in the answer. You ask: Is the first digit less than the number you divide by?
- The answer is yes, so you underline two digits: $3\overline{)\underline{18}18}$
- Then you write the answer for the underlined part: $\begin{array}{r} 6\phantom{18} \\ 3\overline{)\underline{18}18} \end{array}$
- Now you go to the next digit and say the problem for that digit. That digit is less than 3. So you have to underline two digits. **But you must remember to write a digit in the answer. That digit is zero.** $\begin{array}{r} 60\phantom{8} \\ 3\overline{)\underline{18}\,\underline{18}} \end{array}$
- Now you work the problem for the second underlined part. That's the same as the first two digits. You write the answer above the last digit of 18. $\begin{array}{r} 606 \\ 3\overline{)\underline{18}\,\underline{18}} \end{array}$

# Independent Work

## Part 4

**Copy each problem and complete the equation.**

a. $5 = \frac{\phantom{0}}{100}$  b. $5 = \frac{\phantom{0}}{13}$  c. $5 = \frac{\phantom{0}}{26}$

## Part 5

**Write the answer to each problem. Do not copy the problem.**

a. $3\overline{)21}$  d. $3\overline{)0}$  g. $3\overline{)18}$  j. $4\overline{)32}$

b. $4\overline{)24}$  e. $3\overline{)27}$  h. $3\overline{)24}$  k. $4\overline{)4}$

c. $4\overline{)20}$  f. $4\overline{)28}$  i. $4\overline{)0}$

## Part 6

**Copy each problem and work it.**

a. $\frac{19}{3} - 2$  b. $8 + \frac{2}{5}$  c. $\frac{2}{9} + 9$  d. $\frac{15}{4} - 2$

## Part 7

**Write the equation to show the fractions that equal the first fraction.**

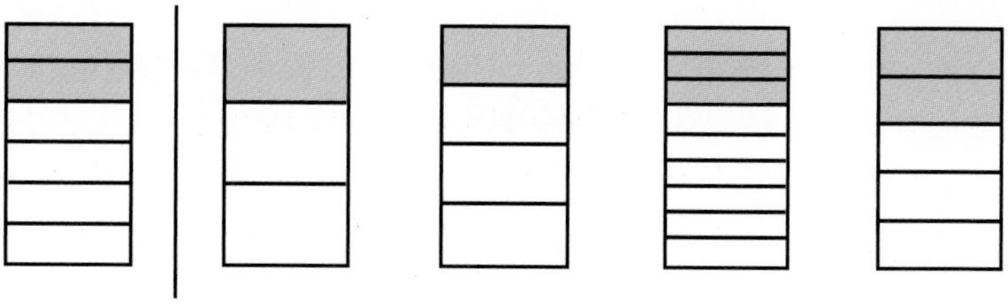

## Part 8 Copy the table. Figure out the missing numbers. Answer the questions.

This table is supposed to show the number of old coins and new coins that two girls collected.

| | Ann | Jan | Total for both girls |
|---|---|---|---|
| Old coins | 16 | | 250 |
| New coins | 240 | | 250 |
| Total coins | | | |

***Questions***

a. Which girl had the larger number of old coins?

b. Which girl had the fewer new coins?

c. What was the total for all the coins?

d. How many total coins did Jan have?

e. How many total new coins were there?

## Part 9 For each problem, make a number family and answer the question.

a. There were 23 more people on the boat than there were on the bus. There were 88 people on the bus. How many were on the boat?

b. A goat ate 184 ounces of food. A rabbit ate 11 ounces of food. How many fewer ounces of food did the rabbit eat than the goat ate?

c. A pine tree was 45 feet taller than a maple tree. The maple tree was 79 feet tall. How tall was the pine tree?

d. Joe threw a ball 145 feet. Jim threw the ball 190 feet. How many feet farther did Jim throw than Joe threw?

# Test 5

## Part 6

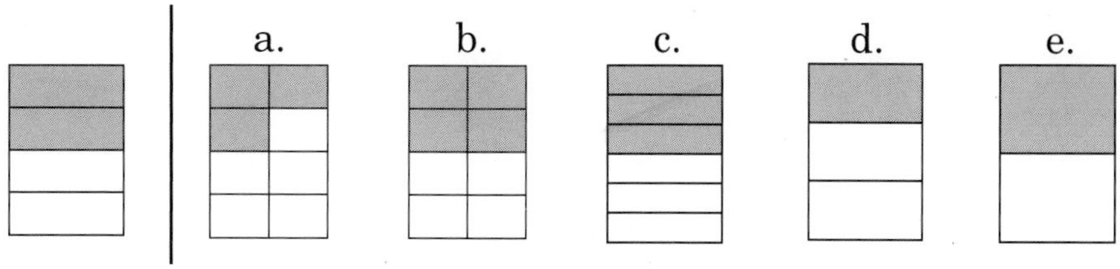


## Part 7

a. $3 = \frac{\quad}{36}$

b. $\frac{\quad}{6} = 24$

c. $\frac{27}{3} =$

## Part 8

a. Tom picked 138 bushels of apples. Ginger picked 127 more bushels than Tom picked. How many bushels of apples did Ginger pick?

b. The car went 63 miles per hour. The tractor went 21 miles per hour. What was the difference in the speed of the vehicles?

c. A box weighed 276 ounces. A jar weighed 19 ounces. How much lighter was the jar than the box?

# Lesson 50

## Part 1

| | | |
|---|---|---|
| __ | 1__ | 2__ |
| 2__ | 35 | 4__ |
| 4__ | 5__ | 6__ |
| 70 | | |

## Part 2

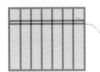

a. $2\overline{)14612}$ b. $2\overline{)6180}$ c. $5\overline{)540}$

When I drop this weight, we'll be able to see if it weighs more than you do.

# Lesson 51

## Part 1

a. 4)24036  b. 5)535  c. 3)2715  d. 5)3010

## Part 2

a. 607 x 53

b. 291 x 86

c. 423 x 78

## Part 3

- You've worked tables with rows that work like number families.
- Here's a new type of table that has number-family rows:
- The numbers in the **first column** tell about the time people **left home.**

  The numbers in the **middle column** show the **time** their trips took.

  The numbers in the **last column** show the time they **arrived downtown**.

**Table for Trips Downtown**

| | Left home | Time of trip | Arrived downtown |
|---|---|---|---|
| Bill | | 1:18 | 8:32 |
| Donna | 7:05 | | 7:43 |
| Ginger | 10:28 | :27 | |

The time Bill **left** is the missing small number in the first row. To find that number, you subtract:

$$\begin{array}{r} 8:\overset{2}{\not{3}}{}^{1}2 \\ -\,1:1\,8 \\ \hline 7:1\,4 \end{array}$$

**Table for Trips Downtown**

| | Left home | Trip downtown | Arrived downtown |
|---|---|---|---|
| a. Hank | 8:29 | | 9:38 |
| b. Ginger | 5:07 | 2:19 | |
| c. Ann | | 1:34 | 8:51 |

## Part 4

a. $\frac{3}{4} \times \frac{8}{4} = \frac{\square}{\square}$  b. $\frac{3}{4} + \frac{8}{4} = \frac{\square}{\square}$  c. $\frac{7}{2} - \frac{3}{2} = \frac{\square}{\square}$

d. $\frac{8}{2} \times \frac{3}{5} = \frac{\square}{\square}$  e. $\frac{1}{6} + \frac{4}{6} = \frac{\square}{\square}$  f. $\frac{4}{3} \times \frac{5}{2} = \frac{\square}{\square}$

## Part 5

| | | |
|---|---|---|
| 7 | 14 | 21 |
| 28 | 35 | 42 |
| 49 | 56 | 63 |
| 70 | | |

# Independent Work

## Part 6

**Write the equation to show the fraction for each mixed number.**

a. $7 + \frac{1}{8}$  b. $3 + \frac{1}{3}$  c. $5 + \frac{7}{9}$

## Part 7

**Make a number family for each problem and figure out the answer.**

a. 67 birds were on the pond. The rest were flying over the pond. If there were 340 birds, how many were flying?

b. The ice cream person sold 34 ice cream cones and 176 ice cream bars. How many ice cream bars and ice cream cones did she sell?

c. The cost of heating a house was $24 in March and $29 in April. How much less did it cost to heat the house during April than in March?

d. The cost of heating a large building in March was $371 more than the cost for April. If the cost in March was $608, what was the cost in April?

## Part 8 Answer the questions. Make a number family for c and d.

This table shows the number of fish and frogs in Hamm's Pond.

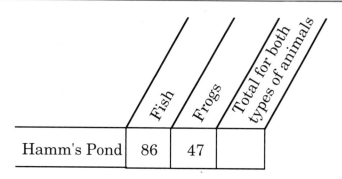

| | Fish | Frogs | Total for both types of animals |
|---|---|---|---|
| Hamm's Pond | 86 | 47 | |

### *Questions*

a. What's the total for both types of animals?

b. Are there more fish or frogs?

c. There are 134 more water bugs in Hamm's Pond than there are fish. How many water bugs are in Hamm's Pond?

d. How many more fish are there than frogs? (**Hint:** Make a number family and figure out the difference number.)

## Part 9 Write the addition problem and the answer for each item.

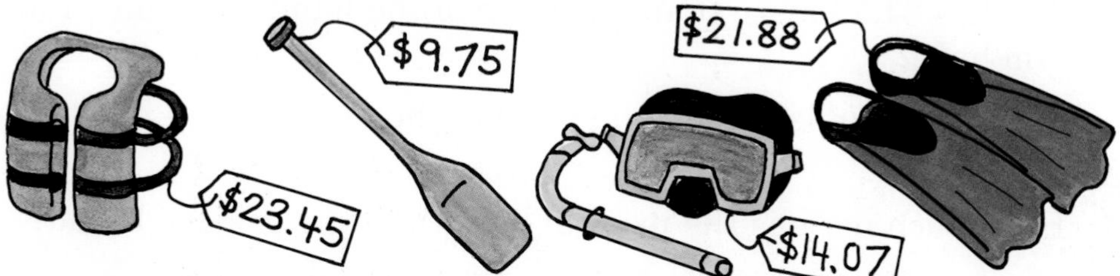


a. Jill buys the life vest and the paddle. How much does she spend?

b. Jim buys the paddle, the snorkle mask and the swim fins. How much does he spend?

c. Ginger buys the life vest and the swim fins. How much does she spend?

# Lesson 52

## Part 1

| | | |
|---|---|---|
| ___ | ___ | 21 |
| ___ | 35 | ___ |
| 49 | ___ | ___ |
| ___ | | |

## Part 2

a. $\frac{5}{3} \times \frac{2}{3} = \square$

b. $\frac{5}{3} - \frac{2}{3} = \square$

c. $\frac{5}{3} + \frac{2}{3} = \square$

d. $\frac{1}{6} + \frac{3}{6} = \square$

e. $\frac{5}{2} \times \frac{2}{2} = \square$

## Part 3

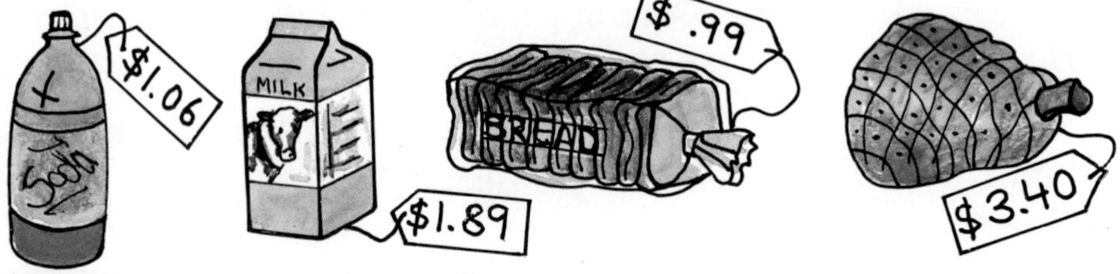


a. How much less is the price of soda pop than the price of milk?

b. Jim has $1.75 more than the price of the ham. How much money does Jim have?

c. Sally has $.26 less than the price of the bread. How much money does Sally have?

d. How much more is the price of ham than the price of bread?

## Part 4

a. $18 + \frac{2}{3}$

b. $5 + \frac{3}{24}$

c. $14 + \frac{6}{7}$

## Part 5

This table is supposed to show the time people left their houses for Rossi's farm, the time of their trips and the time they arrived.

| | Time left | Time of trip | Time arrived |
|---|---|---|---|
| a. Tim | 2:10 | 2:09 | |
| b. Tina | | 3:19 | 4:49 |
| c. Bob | 1:13 | | 4:26 |

# Independent Work

## Part 6

**Write the time shown on each clock.**

a.

b.

c.

## Part 7

**Make a number family and answer the question.**

Joan had $25.86 in the bank. The rest of her money was not in the bank. Joan had $97.59 in all. How much was not in the bank?

## Part 8

**Copy each problem and work it.**

a. 179 x 43

b. 504 x 61

c. 3094 x 27

## Part J

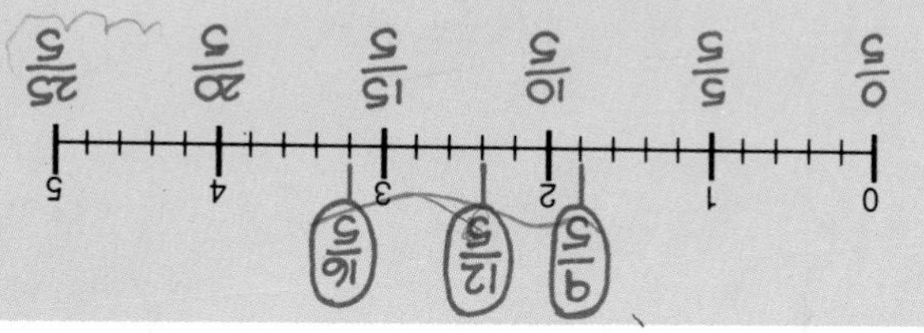

# Lesson 53

## Part 1

a. One of the items costs $3.78 more than the umbrella costs. What is the name of that item?

b. One of the items costs $5.90 less than the fishing pole. What is the name of that item?

c. The swimsuit costs $4.02 more than one item. What is the name of that item?

d. The boots cost $8.02 less than one of the items. Which item?

## Part 2

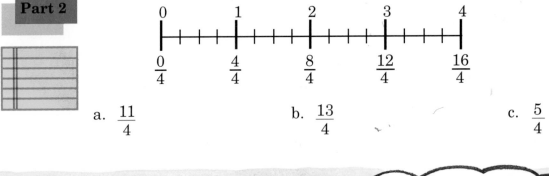

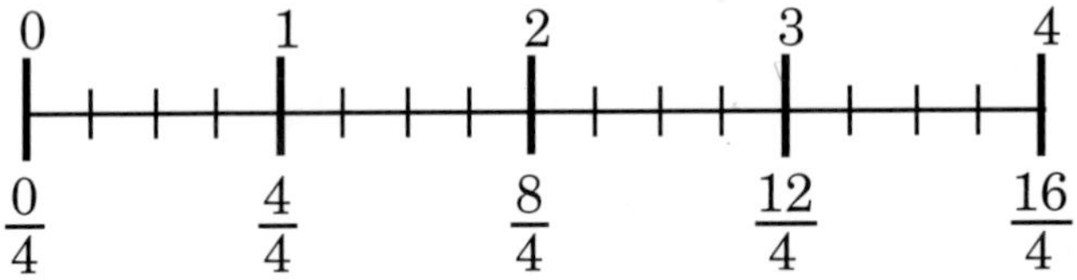


a. $\frac{11}{4}$

b. $\frac{13}{4}$

c. $\frac{5}{4}$

Alex, I said make a fraction that is more than 1. I didn't say make more than one fraction.
5/9 1/3

## Part 3

This table is supposed to show the number of green and red cars that are on Jim's lot and Al's lot.

| | Green cars | Red cars | Total for both colors |
|---|---|---|---|
| Jim's lot | | | |
| Al's lot | 18 | | |
| Total for both lots | | 54 | |

**Fact 1:** There are 23 red cars on Al's lot.

**Fact 2:** The total green cars on both lots is 30.

**Fact 3:** The total for both colors on Jim's lot is 43.

***Questions***

a. Are there fewer green cars or red cars on both lots?

b. How many green and red cars are on Al's lot?

c. There are 31 cars of some color on Jim's lot. What color cars?

d. Are there more green cars on Jim's lot or Al's lot?

# Independent Work

## Part 4

**Copy each problem and figure out the answer.**

a. $\frac{14}{3} + \frac{5}{3} =$  b. $\frac{27}{2} - \frac{3}{2} =$  c. $\frac{50}{3} \times \frac{2}{5} =$

d. $\frac{12}{5} + \frac{12}{5} =$  e. $\frac{2}{7} \times \frac{9}{7} =$  f. $\frac{4}{9} \times \frac{3}{3} =$

## Part 5

**Write the division problem and the answer for each multiplication problem.**

a. 4 x ■ = 36  b. 4 x ■ = 8  c. 4 x ■ = 12

d. 4 x ■ = 24  e. 4 x ■ = 4

## Part 6

**Write the complete equation.**

$$7 = \frac{\quad}{5} = \frac{\quad}{1} = \frac{\quad}{9}$$

## Part 7

**Write the answer to each problem.**

a. $6 \times 7 = ■$ b. $8 \times 7 = ■$ c. $9 \times 7 = ■$ d. $4 \times 7 = ■$

e. $7 \times 7 = ■$ f. $5 \times 7 = ■$ g. $3 \times 7 = ■$

## Part 8

**Write the whole number each fraction equals.**

a. $\frac{6}{1}$ b. $\frac{8}{8}$ c. $\frac{20}{5}$ d. $\frac{20}{4}$

## Part 9

**For each problem, make a number family and answer the question.**

a. In the dance club, there were 45 experienced dancers. The rest were beginners. There were 74 dancers in the club. How many were beginners?

b. Just before dinner, some of the people in the neighborhood were working. 62 people were not working. There was a total of 89 people in the neighborhood. How many were working?

## Part 10

**Copy each problem and work it. Check your work with a calculator.**

a. $\begin{array}{r} 390 \\ \times\ 94 \\ \hline \end{array}$ b. $\begin{array}{r} 194 \\ \times\ 93 \\ \hline \end{array}$ c. $\begin{array}{r} 406 \\ \times\ 5 \\ \hline \end{array}$

# Lesson 54

## Part 1

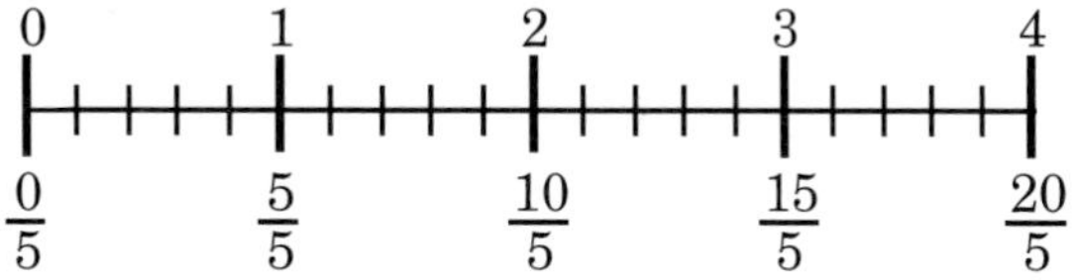


a. $\frac{2}{5}$  b. $\frac{19}{5}$  c. $\frac{13}{5}$

## Part 2

a. $\frac{2}{3} \times \frac{3}{2} =$  b. $\frac{1}{10} \times \frac{3}{2} =$  c. $\frac{8}{3} \times 2 =$

d. $4 \times \frac{2}{9} =$  e. $\frac{10}{3} \times \frac{2}{2} =$

## Part 3

- If you multiply any value by a fraction that equals 1, the answer equals the value you started with.
- Here's 2 times 1 equals 2: **$2 \times 1 = 2$**
  So 2 times **any fraction that equals 1** equals 2.
- Here's 2 times a fraction that equals 1: $2 \times \frac{5}{5} = 2$
- First, we change 2 into a fraction that equals 2: $\frac{2}{1} \times \frac{5}{5} = \boxed{\frac{\phantom{0}}{\phantom{0}}}$
- Then we multiply. When we multiply on top, we get **10.** $\frac{2}{1} \times \frac{5}{5} = \boxed{\frac{10}{\phantom{0}}}$
- When we multiply on the bottom, we get **5.** $\frac{2}{1} \times \frac{5}{5} = \boxed{\frac{10}{5}}$
- Here's a picture of $\frac{10}{5}$. The picture shows that $\frac{10}{5}$ equals 2 whole units.

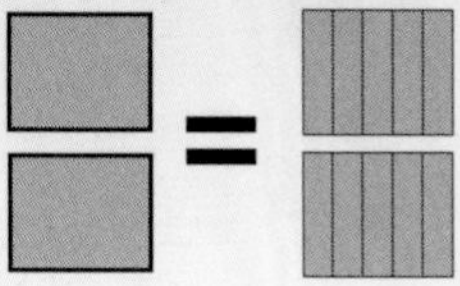

a. $5 \times \frac{2}{2} =$  b. $\frac{3}{3} \times 9 =$  c. $\frac{7}{7} \times 3 =$  d. $8 \times \frac{4}{4} =$

## Part 4

a. The bike costs $14.89 more than one of the items. Which item?

b. Jim has $13.24 more than the cost of the surfboard. How much money does Jim have?

c. How much less is the price of the skateboard than the price of the raft?

d. Jan wants to buy an item that costs $59.81 less than the price of the surfboard. Which item does Jan want to buy?

e. How much more is the price of the bike than the price of the swim fins?

## Independent Work

## Part 5

**Use the completed table from part 1 of your workbook. Answer the questions.**

a. How many boys went to Washington and Roosevelt Schools?

b. Did fewer girls go to Washington School or Roosevelt School?

c. Were there more boys in both schools or more girls?

d. How many children went to Roosevelt School?

## Part 6

**Write the answer to each problem.**

| | | |
|---|---|---|
| 1 | 2 | 3 |
| 4 | 5 | 6 |
| 7 | 8 | 9 |
| 10 | | |

a. 7 x 5 = ■
b. 7 x 1 = ■
c. 7 x 3 = ■
d. 7 x 7 = ■
e. 7 x 9 = ■
f. 7 x 2 = ■
g. 7 x 0 = ■
h. 7 x 6 = ■
i. 7 x 8 = ■
j. 7 x 4 = ■
k. 7 x 10 = ■

## Part 7 Copy each problem and work it.

a. $\frac{4}{3} \times \frac{2}{3} =$

b. $\frac{12}{5} - \frac{12}{5} =$

c. $\frac{3}{4} \times \frac{7}{4} =$

d. $\frac{7}{8} \times \frac{4}{1} =$

e. $\frac{1}{32} + \frac{54}{32} =$

f. $\frac{203}{4} - \frac{200}{4} =$

## Part 8

a. 86 x 52

b. 145 x 32

## Part J

b.

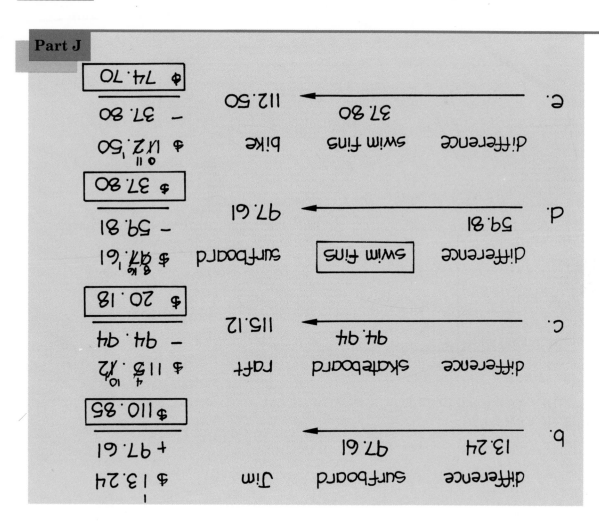

| difference | surfboard | Jim |
|---|---|---|
| 13.24 | 97.61 | → |

$13.24 + $97.61 = **$110.85**

c.

| difference | skateboard | raft |
|---|---|---|
| | 94.94 | → 115.12 |

$115.12 − 94.94 = **$20.18**

d.

| difference | swim fins | surfboard |
|---|---|---|
| 59.81 | | → 97.61 |

$97.61 − 59.81 = **$37.80**

e.

| difference | swim fins | bike |
|---|---|---|
| | 37.80 | → 112.50 |

$112.50 − 37.80 = **$74.70**

# Lesson 55

## Part 1

a. 7 x 9

b. 7 x 10

c. 7 x 7

d. 7 x 8

e. 7 x 3

f. 7 x 4

g. 7 x 6

## Part 2

a. $10 \times \frac{3}{5} =$  b. $\frac{2}{8} \times \frac{1}{3} =$  c. $\frac{5}{4} \times \frac{7}{5} =$  d. $\frac{4}{5} \times 3 =$

## Part 3

This table is supposed to show the number of tall trees and trees that are not tall in Mountain Park and River Park.

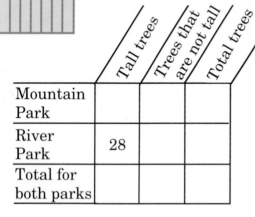

| | Tall trees | Trees that are not tall | Total trees |
|---|---|---|---|
| Mountain Park | | | |
| River Park | 28 | | |
| Total for both parks | | | |

***Facts***

1. The total number of trees in River Park is 112.
2. In Mountain Park, 73 trees are not tall.
3. The total number of trees in Mountain Park is 175.

***Questions***

a. In which park are there fewer trees that are not tall, Mountain Park or River Park?

b. How many tall trees are there in both parks?

c. Are there more trees in Mountain Park or in River Park?

## Part 4

a. $\frac{2}{7} \times \frac{4}{4} =$  b. $5 \times \frac{9}{9} =$  c. $\frac{8}{5} \times \frac{3}{3} =$

d. $3 \times \frac{6}{6} =$  e. $2 \times \frac{8}{8} =$

# Independent Work

## Part 5

**Figure out the area of each rectangle.**

a.

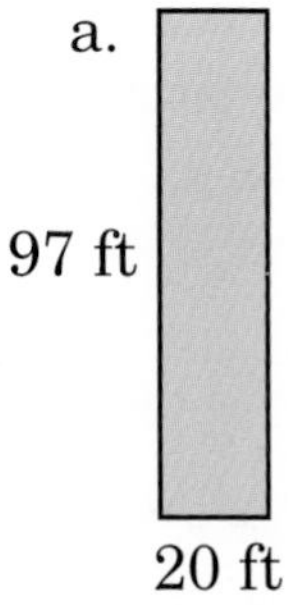


b.

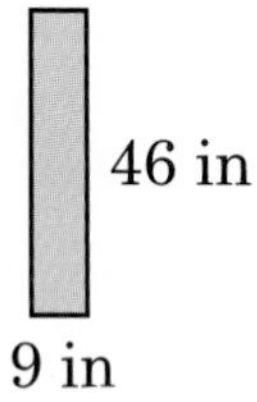


## Part 6

**Copy the table. Figure out the missing time for each row. Then answer the questions.**

This table shows the times of the trips that different groups took.

| | Leave | Time of trip | Arrived |
|---|---|---|---|
| Cheerleaders | 2:30 | | 4:46 |
| Glee Club | | :46 | 3:55 |
| Swim Team | 6:10 | 1:18 | |

***Questions***

a. When did the swim team arrive?

b. How long was the cheerleaders' trip?

c. When did the Glee Club leave?

## Part 7

**For each mixed number, write the complete equation.**

a. $876 + \frac{7}{8} =$  b. $4 + \frac{8}{150} =$

## Part 8

**Write the multiplication or addition fact for each set of numbers.**

a. 1, 1, 2  b. 10, 1, 10  c. 1, 1, 1

## Part 9

**Write the equation that shows all the fractions that equal the fraction for picture a.**

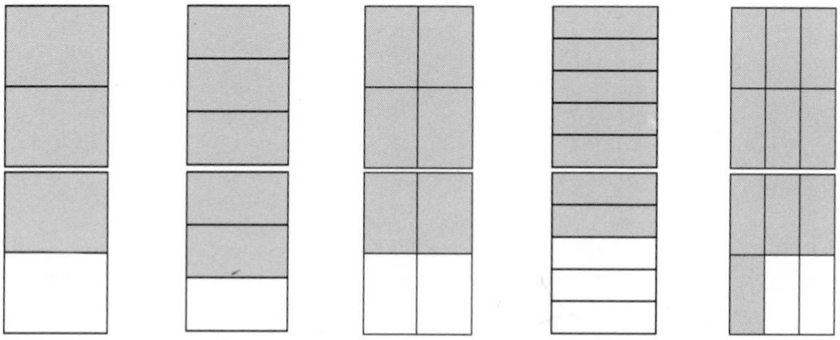

a. b. c. d. e.

## Part 10

**Copy each problem and write the fraction it equals.**

a. $3 - \frac{7}{5} =$  b. $\frac{3}{7} - \frac{3}{7} =$  c. $\frac{20}{4} + \frac{40}{4} =$  d. $\frac{11}{8} + 2 =$

## Part 11

**Copy each item and complete the equation.**

a. 58 = 9 x ■ + ■
■

b. 30 = 9 x ■ + ■
■

c. 23 = 9 x ■ + ■
■

d. 71 = 9 x ■ + ■
■

# Lesson 56

## Part 1

a. $\frac{14}{7}$ b. $\frac{54}{9}$ c. $\frac{45}{5}$ d. $\frac{30}{5}$

## Part 2

a. $\frac{\quad}{3} = 6$ b. $\frac{35}{7} =$ c. $\frac{\quad}{9} = 6$ d. $\frac{6}{2} =$

## Part 3

a. $4\overline{)8}$ b. $4\overline{)16}$ c. $4\overline{)24}$ d. $4\overline{)12}$ e. $4\overline{)36}$

## Part 4

a. 4 x 8 = ■ b. 7 x 6 = ■ c. 4 x 6 = ■ d. 7 x 9 = ■

e. 7 x 4 = ■ f. 7 x 8 = ■ g. 4 x 4 = ■ h. 4 x 7 = ■

## Part 5

This table is supposed to show the number of hamburgers and hot dogs eaten by boys and girls at Winfrey School.

| | Hamburgers | Hot dogs | Total for both sandwiches |
|---|---|---|---|
| Boys | | | |
| Girls | | | |
| Total students | | 233 | |

***Facts***

1. The girls at Winfrey School ate 213 sandwiches.
2. Students ate 414 hamburgers.
3. The girls ate 87 hot dogs.

***Questions***

a. How many sandwiches did the students at Winfrey School eat?

b. Did the girls eat fewer hamburgers or fewer hot dogs?

c. How many hot dogs did the boys eat?

## Independent Work

## Part 6

**Write each fraction as a division problem and write the answer. The top number is the number under the division sign.**

a. $\frac{14}{7}$ b. $\frac{54}{9}$ c. $\frac{45}{5}$ d. $\frac{18}{3}$

e. $\frac{8}{2}$ f. $\frac{60}{6}$ g. $\frac{8}{4}$ h. $\frac{27}{3}$

**Part 7** **Copy each item. Use the number line to write the mixed number and complete each equation.**

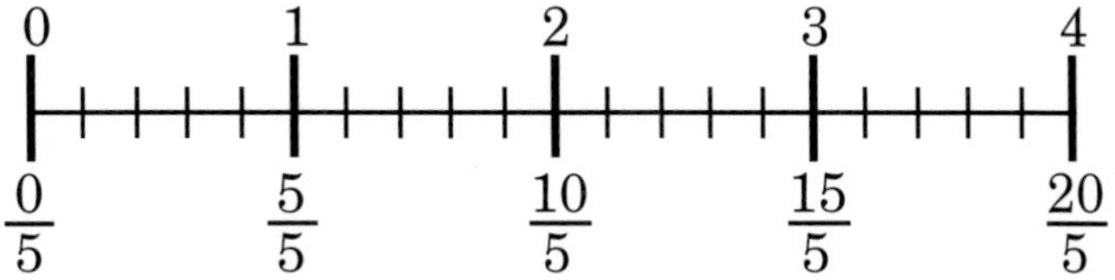


a. $\frac{18}{5}$ b. $\frac{4}{5}$ c. $\frac{11}{5}$

**Part 8** **Write the fraction for each description.**

a. The top number of the fraction is 34. The fraction equals 1 whole.

b. The picture for this fraction shows 13 parts shaded. There are 17 parts in each group.

c. The picture for this fraction has 7 parts in each group. 7 parts are shaded.

d. The numbers for this fraction are 16 and 13. The fraction is more than 1 whole.

e. The numbers for this fraction are 1 and 6. The fraction is less than 1 whole.

**Part 9** **Copy each problem and complete the equation.**

a. $\frac{15}{7} + \frac{27}{7} =$ b. $\frac{3}{10} \times \frac{1}{4} =$ c. $\frac{4}{9} \times \frac{26}{5} =$ d. $\frac{5}{37} - \frac{5}{37} =$

## Part 10

**Make the number families. Write the number problems and the answers.**

a. Joan has $1.25 less than the cost of a steak. How much money does Joan have?

b. The chicken sandwich costs $1.30 more than one of the items. Which item?

## Part 11

**Copy the problems and work them. Check your work with a calculator.**

a. $\begin{array}{r} 584 \\ \underline{\times \; 20} \end{array}$

b. $\begin{array}{r} 56 \\ \underline{\times \; 35} \end{array}$

c. $\begin{array}{r} 1503 \\ \underline{\times \quad 9} \end{array}$

## Part 12

**For each problem, make the number family. Figure out the missing number.**

| | Hamburgers |
|---|---|
| Boys | 278 |
| Girls | 126 |
| Total students | 404 |

***Questions***

a. The number of girls who ate hamburgers is 126 less than the number of girls who ate melons. How many girls ate melons?

b. The number of boys who ate vegetables is 45 more than the number of boys who ate hamburgers. How many boys ate vegetables?

# Lesson 57

## Part 1

a. 6 x 3 = ■ b. 5 x 7 = ■ c. 7 x 4 = ■ d. 6 x 7 = ■

e. 8 x 7 = ■ f. 8 x 3 = ■ g. 7 x 3 = ■

## Part 2

This table is supposed to show the number of apples that were ripe and the number of apples that were not ripe in Vista Orchard and Brandon Orchard.

| | Ripe | Not ripe | Total apples |
|---|---|---|---|
| Vista Orchard | | | |
| Brandon Orchard | | | |
| Total for both orchards | | | 4284 |

***Facts***

1. A total of 875 apples were ripe.
2. In Vista Orchard, 1064 apples were not ripe.
3. In Vista Orchard, 851 apples were ripe.

***Questions***

a. Were there fewer ripe apples in Vista Orchard or in Brandon Orchard?

b. Which orchard had more apples that were not ripe?

c. How many apples that were not ripe were in both orchards?

d. Which orchard had more total apples?

## Part 3

a. $\frac{24}{3}$ b. $\frac{16}{2}$ c. $\frac{36}{9}$ d. $\frac{24}{4}$

e. $\frac{18}{3}$ f. $\frac{18}{9}$ g. $\frac{8}{8}$ h. $\frac{12}{3}$

## Independent Work

### Part 4

**Copy each problem and work it.**

a. $3\overline{)96}$ b. $4\overline{)8284}$ c. $3\overline{)1824}$

### Part 5

**Copy each problem and work it. Check your work with a calculator.**

a. $\begin{array}{r} 476 \\ \underline{\times \quad 3} \end{array}$ b. $\begin{array}{r} 476 \\ \underline{\times \quad 5} \end{array}$ c. $\begin{array}{r} 476 \\ \underline{\times \quad 2} \end{array}$ d. $\begin{array}{r} 76 \\ \underline{\times 49} \end{array}$ e. $\begin{array}{r} 476 \\ \underline{\times \quad 0} \end{array}$

### Part 6

**Make a number family for each problem. Answer the questions.**

a. There were butterflies and moths in a jar. There were 37 moths in the jar. There were 53 insects in all. How many butterflies were in the jar?

b. There were moths and butterflies in a jar. There were 37 fewer moths than butterflies. There were 53 moths. How many butterflies were there?

c. A hippo weighed 6480 pounds. A giraffe weighed 4020 pounds. How much heavier is the hippo than the giraffe?

d. Jim had $297 in all. If he had $175 in the bank, how much money was not in the bank?

### Part 7

**Copy each problem and work it.**

a. $\frac{5}{7} \times \frac{5}{7} =$ b. $\frac{5}{7} - \frac{5}{7} =$ c. $\frac{5}{7} + \frac{5}{7} =$

d. $\frac{8}{3} \times \frac{4}{1} =$ e. $\frac{7}{2} - \frac{4}{2} =$ f. $\frac{4}{10} \times \frac{5}{2} =$

## Part 8

**Copy each problem and complete the equation to show the fraction and the whole number it equals.**

a. $\frac{7}{1} =$

b. $\frac{\phantom{0}}{3} = 9$

c. $\frac{0}{9} =$

d. $\frac{12}{4} =$

e. $\frac{\phantom{0}}{2} = 10$

f. $\frac{\phantom{0}}{4} = 8$

g. $\frac{5}{5} =$

## Part 9

**Write the complete equation.**

a. $27 + \frac{3}{5} =$

b. $8 + \frac{7}{143} =$

## Part 10

**Copy each problem and work it.**

a. $4\overline{)28}$

b. $4\overline{)32}$

c. $3\overline{)24}$

d. $3\overline{)18}$

e. $4\overline{)12}$

f. $4\overline{)24}$

g. $4\overline{)36}$

h. $4\overline{)32}$

i. $3\overline{)27}$

We'll put 5 pigs in each pen. We have 45 pigs. So you know what that means.
??

# Lesson 58

## Part 1

a. $\frac{12}{4}$ b. $\frac{20}{5}$ c. $\frac{63}{9}$ d. $\frac{12}{2}$

e. $\frac{21}{3}$ f. $\frac{5}{5}$ g. $\frac{27}{9}$ h. $\frac{7}{1}$

## Part 2

a. $4 \times \frac{6}{6} =$ b. $\frac{2}{3} \times \frac{9}{7} =$ c. $\frac{5}{9} \times \frac{4}{4} =$ d. $5 \times \frac{8}{8} =$

e. $\frac{12}{4} \times \frac{5}{10} =$ f. $\frac{7}{2} \times \frac{2}{7} =$ g. $\frac{1}{2} \times \frac{7}{7} =$

## Part 3

- These are special multiplication facts. These facts are special because you can make a square for each fact.
- The number in each row is exactly the same as the number in each column.

1 x 1 = 1
2 x 2 = 4
3 x 3 = 9
4 x 4 = 16
5 x 5 = 25
6 x 6 = 36
7 x 7 = 49
8 x 8 = 64
9 x 9 = 81

- Here's 5 times 5: 5 x 5 =
- There are 5 in each row and 5 in each column.
- There are 25 square units in all: 5 x 5 = 25

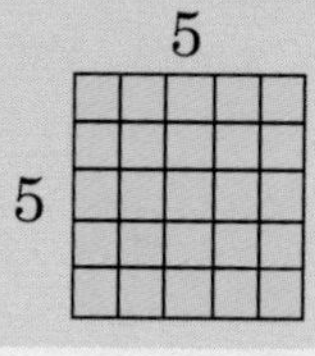


- These are **new** multiplication facts for squares:

6 x 6 = 36
8 x 8 = 64

a. 7 x 7 = ■ d. 5 x 5 = ■ g. 3 x 3 = ■

b. 9 x 9 = ■ e. 4 x 4 = ■ h. 1 x 1 = ■

c. 6 x 6 = ■ f. 2 x 2 = ■ i. 8 x 8 = ■

**Part 4**

a. $\frac{29}{3} = 9 +$

b. $\frac{48}{9} = 5 +$

c. $\frac{70}{8} = 8 +$

d. $\frac{39}{5} = 7 +$

e. $\frac{22}{3} = 7 +$

## Independent Work

**Part 5** **For each problem, make a number family and answer the question.**

a. Tim has $2.37 less than the price of the radio. How much money does Tim have?

b. The price of one item is $12.46 more than the price of the note pad. Which item is that?

c. Janice has $5.67 more than the price of the ruler. How much money does Janice have?

**Part 6** **Copy each problem and work it.**

a. $9\overline{)36630}$

b. $2\overline{)68}$

c. $4\overline{)8244}$

## Part 7 Figure out the area of each rectangle.

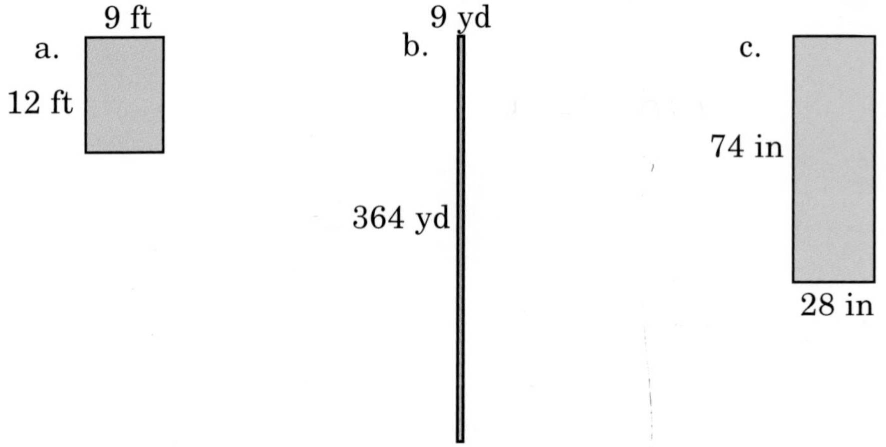


## Part 8 Copy and complete each equation.

a. $9 = \frac{\quad}{1} = \frac{\quad}{6}$

b. $\frac{9}{3} =$

c. $9 = \frac{\quad}{4} = \frac{\quad}{7}$

d. $5 = \frac{\quad}{6}$

e. $\frac{15}{5} =$

## Part 9 Write the fraction for each description.

a. The fraction is less than 1. The numbers are 68 and 29.

b. The picture shows 7 parts in each unit. There are 20 parts shaded.

c. The numbers are 10 and 2. The fraction is more than 1.

d. The picture shows 72 parts shaded. There are 100 parts in each unit.

**Part 10** **Write the division problem and the answer for each multiplication problem.**

a. $4 \times \blacksquare = 32$ b. $9 \times \blacksquare = 27$ c. $7 \times \blacksquare = 49$ d. $4 \times \blacksquare = 16$

**Part 11** **Make a number family for each problem and write the answer.**

a. Debbie had money in her purse and in her pocket. She had \$3.45 in her pocket and \$18.97 in all. How much money did Debbie have in her purse?

b. A club earned money by washing cars and by selling wood. The club earned \$195 washing cars and \$72 selling wood. How much money did the club earn in all?

# Lesson 59

## Part 1

a. $\frac{450}{5}$    b. $\frac{243}{3}$    c. $\frac{4120}{2}$    d. $\frac{6240}{3}$

## Part 2

a. Fran had more money than the price of the wallet. After Fran bought the wallet, she still had $4.44. How much money did she start out with?

b. Sid didn't have enough money to buy the sunglasses. The sunglasses cost $5.30 more than he had. How much did Sid have?

c. Janice started out with more than the price of the backpack. After she bought the backpack, she still had $34.11. How much did she start out with?

d. Ginger didn't have enough money to buy the watch. She needed $3.45 more to buy the watch. How much money did she have?

## Part 3

| | | |
|---|---|---|
| a. 5 x 5 = ■ | d. 9 x 9 = ■ | g. 8 x 8 = ■ |
| b. 3 x 3 = ■ | e. 7 x 7 = ■ | h. 2 x 2 = ■ |
| c. 6 x 6 = ■ | f. 4 x 4 = ■ | i. 1 x 1 = ■ |

## Part 4

This table is supposed to show the cows and sheep raised on the Double X Ranch and the Bar Z Ranch.

| | Cows | Sheep | Total animals |
|---|---|---|---|
| Double X Ranch | | | |
| Bar Z Ranch | | | |
| Total for both ranches | | | |

***Facts***

1. At Double X Ranch, there are 1462 cows and sheep.
2. The Bar Z Ranch has 620 cows.
3. The Double X Ranch has 531 sheep.
4. The total cows and sheep on Bar Z Ranch is 1531.

***Questions***

a. Which ranch has fewer cows?
b. How many total cows are on both ranches?
c. Which ranch has 911 sheep?
d. Which is greater, the total number of cows or the total number of sheep?

## Independent Work

## Part 5

**Copy each problem and work it.**

a. $3\overline{)27}$
b. $3\overline{)12}$
c. $4\overline{)12}$
d. $3\overline{)24}$
e. $3\overline{)21}$
f. $4\overline{)28}$
g. $4\overline{)24}$
h. $4\overline{)36}$
i. $3\overline{)24}$
j. $4\overline{)16}$
k. $4\overline{)32}$
l. $3\overline{)15}$

## Part 6

**Copy each equation and write the missing numbers.**

a. 37 = 5 x ■ + ■
b. 50 = 7 x ■ + ■
c. 15 = 4 x ■ + ■
d. 60 = 7 x ■ + ■

## Part 7

**Copy the mixed number and write the equation for the fraction it equals.**

a. $6 + \frac{4}{9} =$ b. $8 + \frac{2}{5} =$ c. $37 + \frac{1}{4} =$

## Part 8

**Copy each equation and complete it.**

a. $6 = \frac{}{5}$ b. $4 = \frac{}{3}$ c. $\frac{10}{5} =$ d. $3 = \frac{}{3}$ e. $\frac{3}{3} =$ f. $7 = \frac{}{4}$

## Part 9

**Write the multiplication fact or the addition fact for each group of numbers.**

a. 14, 7, 2 b. 4, 8, 4 c. 50, 54, 4 d. 2, 3, 6 e. 25, 2, 50

## Part 10

**Copy each number line. Write the fraction for the whole numbers.**

a.

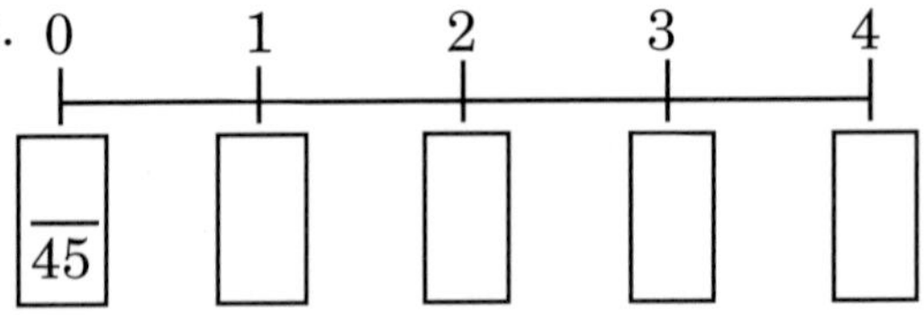


b.

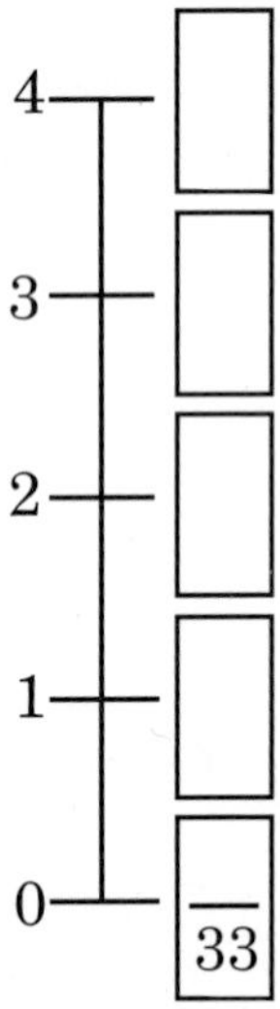

**Part 11** **Copy each problem and work it. Check your work with a calculator.**

a. $\begin{array}{r} 1315 \\ \underline{\text{x} \quad 2} \end{array}$ b. $\begin{array}{r} 23 \\ \underline{\text{x}\ 74} \end{array}$ c. $\begin{array}{r} 905 \\ \underline{+ \quad 9} \end{array}$ d. $\begin{array}{r} 22 \\ \underline{\text{x}\ 53} \end{array}$ e. $\begin{array}{r} 96 \\ \underline{-\ 34} \end{array}$

**Part 12** **Work each problem.**

a. A field is 368 meters long and 70 meters wide. What is the area of the field?

b. A picture is 42 inches high and 24 inches wide. What is the area of the picture?

# Lesson 60

## Part 1

a. $\frac{8}{2}$   b. $\frac{18}{3}$   c. $\frac{35}{5}$   d. $\frac{32}{4}$

## Part 2

| | | |
|---|---|---|
| 1 | 2 | 3 |
| 4 | 5 | 6 |
| 7 | 8 | 9 |
| 10 | | |

a. $7\overline{)56}$   d. $7\overline{)42}$   g. $7\overline{)63}$

b. $7\overline{)14}$   e. $7\overline{)49}$   h. $7\overline{)7}$

c. $7\overline{)28}$   f. $7\overline{)21}$

## Part 3

a. $\frac{3}{4} \times \blacksquare = \frac{18}{24}$   b. $5 \times \blacksquare = \frac{35}{5}$   c. $\frac{9}{2} \times \blacksquare = \frac{36}{4}$

d. $4 \times \blacksquare = \frac{40}{10}$   e. $8 \times \blacksquare = \frac{16}{9}$

## Part 4

a. Jane has $34.86. She wants to buy the backpack. How much change will she get?

b. James has $12.78. He wants to buy the backpack. How much more money does he need?

c. Donna has $5.20. How much more money does she need to buy the sunglasses?

d. Dave had more than the price of the wallet. After Dave bought the wallet, he still had $13.44. How much money did Dave start out with?

# Lesson 61

## Part 1

a. $\frac{3}{7} \times \blacksquare = \frac{6}{14}$

b. $\frac{6}{5} \times \blacksquare = \frac{30}{30}$

c. $\frac{3}{8} \times \blacksquare = \frac{18}{16}$

d. $\frac{5}{1} \times \blacksquare = \frac{20}{4}$

## Part 2

a. $\frac{63}{3}$

b. $\frac{205}{5}$

c. $\frac{36}{9}$

d. $\frac{836}{4}$

e. $\frac{60}{6}$

## Part 3

a. The bike costs $11.14 more than what Jerry has.

b. Donna has $3.45 more than the price of the bike.

c. If Ginger bought the bike, she would still have $45.60.

d. If Larry had $12.00 more than he has, he'd be able to buy the bike.

## Independent Work

## Part 4

**Copy each number line and write the fractions for the whole numbers.**

a.

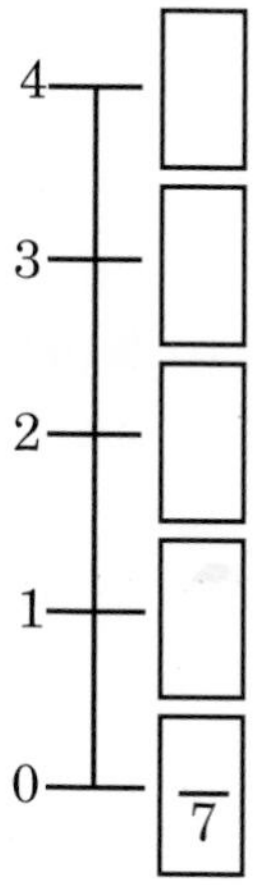


b.

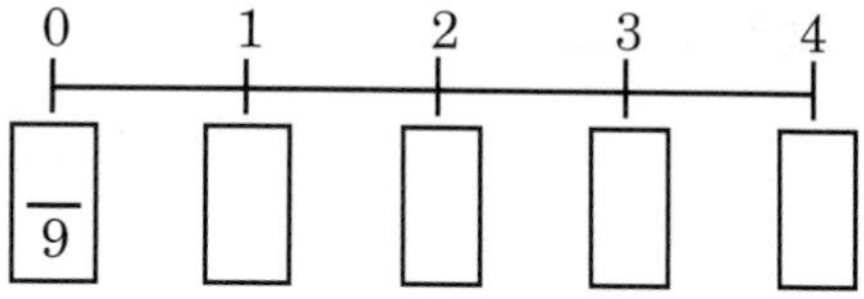

## Part 5

**Copy each problem and write the answer.**

a. 7 x 6 = ■ d. 8 x 7 = ■ g. 6 x 7 = ■

b. 8 x 8 = ■ e. 7 x 4 = ■ h. 7 x 8 = ■

c. 6 x 6 = ■ f. 3 x 7 = ■ i. 7 x 7 = ■

## Part 6

**Copy each equation and complete it.**

a. $4 = \frac{\phantom{0}}{3}$ b. $5 = \frac{\phantom{0}}{3}$ c. $\frac{6}{3} =$

d. $6 = \frac{\phantom{0}}{3}$ e. $\frac{\phantom{0}}{3} = 7$ f. $\frac{10}{2} =$

## Part 7

**Copy each problem and work it. Check your work with a calculator.**

a. $\begin{array}{r} 327 \\ \times\ \ 52 \\ \hline \end{array}$ b. $\begin{array}{r} 91 \\ \times\ 40 \\ \hline \end{array}$ c. $\begin{array}{r} 86 \\ -\ 14 \\ \hline \end{array}$ d. $\begin{array}{r} 53 \\ +\ 75 \\ \hline \end{array}$ e. $\begin{array}{r} 98 \\ \times\ 27 \\ \hline \end{array}$

## Part 8

**Figure out the missing numbers and answer the questions.**

This table shows the times for different people who went to work.

| | Left home at | Time of trip | Arrived at work |
|---|---|---|---|
| Jimbo | 7:09 | | 9:36 |
| Alex | 6:32 | 2:01 | |
| Tina | | 1:48 | 9:51 |

***Questions***

a. How long did it take Jimbo to get to work?

b. At what time did Alex arrive at work?

c. What time did Tina leave home?

# Lesson 62

## Part 1

a. When Larry buys the cooler, he'll still have $13.45. How much money does he have now?

b. Hank had $35.99. Hank bought one of the items. After he bought it, he still had $17.09. Which item did he buy?

c. Steve had $24.38. How much would Steve have left if he bought the folding chair?

d. Jill had $17.09. Jill needed $2.06 to buy one of the items. Which item?

## Part 2

a. $\frac{510}{5}$ b. $\frac{369}{9}$ c. $\frac{804}{4}$ d. $\frac{1624}{4}$

## Independent Work

## Part 3

**Copy each problem and write the answer.**

| | | | |
|---|---|---|---|
| a. 4 x 6 = ■ | d. 9 x 6 = ■ | g. 6 x 2 = ■ | j. 8 x 8 = ■ |
| b. 4 x 7 = ■ | e. 7 x 3 = ■ | h. 7 x 6 = ■ | k. 6 x 6 = ■ |
| c. 6 x 5 = ■ | f. 7 x 8 = ■ | i. 3 x 6 = ■ | l. 7 x 7 = ■ |

## Part 4

**Copy each number line and write the fractions for the whole numbers.**

a. 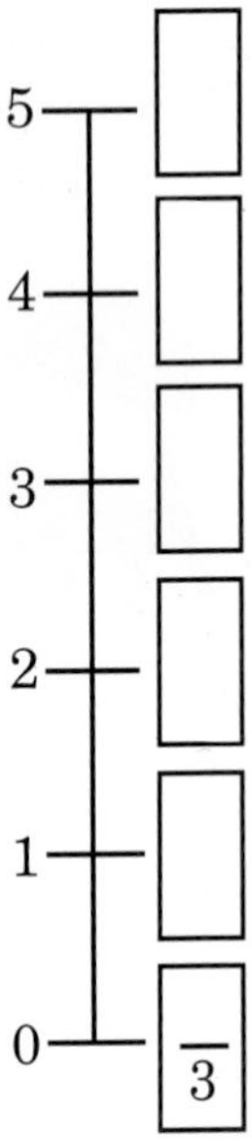


b. 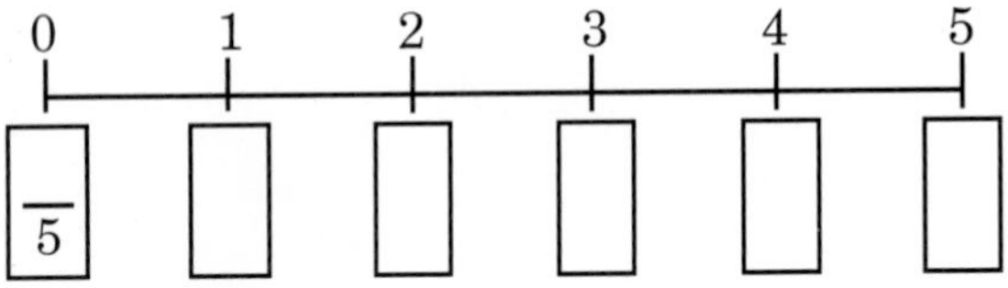


## Part 5

**Make a number family for each problem. Work the problem and box the answer.**

a. 15 students have been to the dentist recently. The rest of the students haven't been to the dentist recently. If there are 139 students, how many have not been to the dentist recently?

b. A phone book had 132 more yellow pages than white pages. There were 389 white pages. How many yellow pages were there?

c. 164 girls were in the race. 231 boys were in the race. How many fewer girls than boys were in the race?

d. Donna had $83.50 in all. She kept $45.03 in a jar. She kept the rest of her money in her wallet. How much did she keep in her wallet?

## Part 6

**Write the number problem and figure out the area for each problem.**

a. 

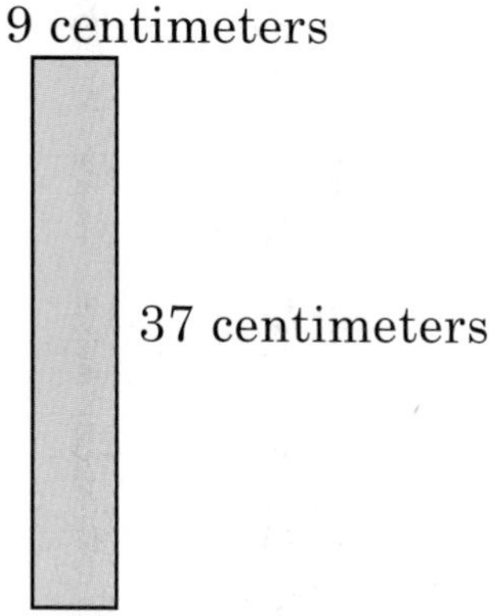


b. A field is 13 yards wide and 67 yards long. What is the area of the field?

## Part 7

**Write the fraction for each item.**

a. 

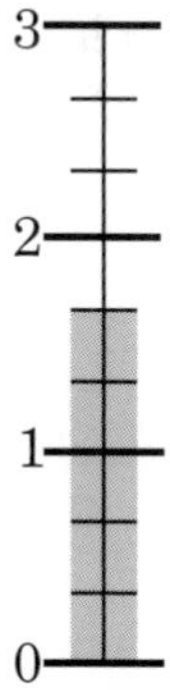


b. 

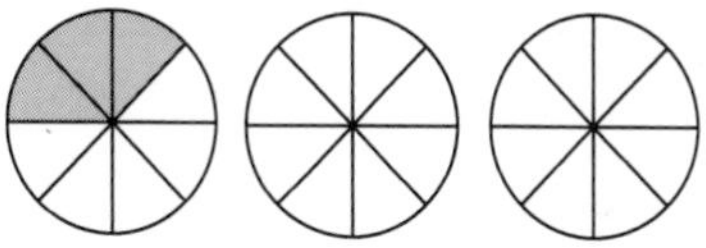

c. 

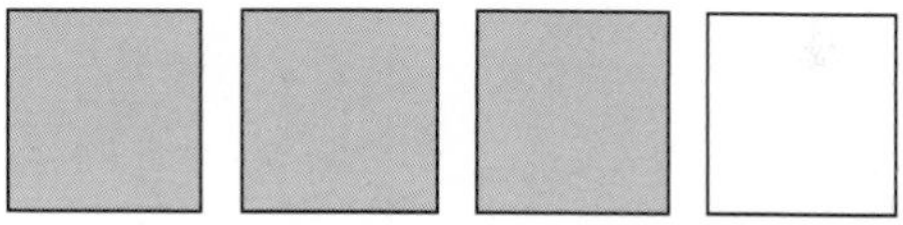

## Part 8

**Write the addition or multiplication fact for each group of numbers. Write each equation so the equal sign comes after the first number.**

a. 27, 4, 31 b. 1, 8, 8 c. 90, 100, 10

## Part 9

**Copy both problems and complete the equation. For the problem that multiplies by 1, write another equation for the the equivalent fractions.**

a. $\frac{7}{4} \times \frac{8}{8} = \blacksquare$ b. $\frac{16}{1} \times \frac{9}{10} = \blacksquare$

# Lesson 63

## Part 1

a. 47 = 7 x ■ + ■ □

b. 65 = 9 x ■ + ■ □

c. 35 = 4 x ■ + ■ □

d. 58 = 9 x ■ + ■ □

e. 31 = 5 x ■ + ■ □

## Part 2

| | |
|---|---|
| 6 | 12 |
| 18 | 24 |
| 30 | 36 |
| 42 | 48 |
| 54 | 60 |

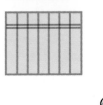

## Part 3

a. Van bought the radio and still had $4.60. How much money did he start out with?

b. Jan needs $13.13 more than she has to buy the radio. How much does Jan have?

c. Dan has $87.07. If he buys the item he wants, he'll have $21.83 left. Which item does Dan want to buy?

d. Fran has $57.12. How much more does she need to buy the watch?

# Independent Work

## Part 4

**Copy each problem and write the answer.**

a. $8 \times 8 = ■$
b. $8 \times 6 = ■$
c. $7 \times 7 = ■$
d. $6 \times 6 = ■$
e. $6 \times 7 = ■$
f. $6 \times 4 = ■$
g. $7 \times 8 = ■$
h. $6 \times 8 = ■$
i. $4 \times 6 = ■$

## Part 5

**Copy each equation and complete it.**

a. $6 = \frac{\phantom{0}}{5}$
b. $\frac{\phantom{0}}{3} = 4$
c. $\frac{\phantom{0}}{5} = 9$
d. $\frac{6}{3} =$
e. $\frac{8}{2} =$
f. $5 = \frac{\phantom{0}}{3}$

## Part 6

**Write the answer for each problem.**

a. $7\overline{)49}$
b. $7\overline{)14}$
c. $7\overline{)28}$
d. $7\overline{)42}$
e. $7\overline{)35}$
f. $7\overline{)56}$
g. $7\overline{)63}$
h. $7\overline{)0}$
i. $7\overline{)21}$

## Part 7

**Copy the mixed number and write the fraction addition equation.**

a. $7 + \frac{5}{4}$
b. $4 + \frac{11}{306}$
c. $35 + \frac{1}{2}$

## Part 8

**For items a–d, write the fraction. For items e–h, write complete equations.**

a. 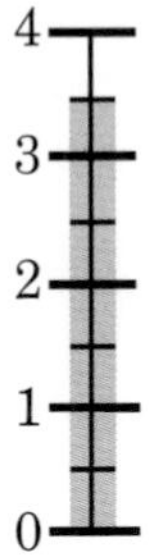

b. 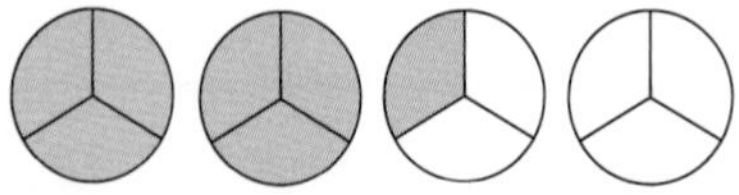

c. The numbers are 42 and 37. The fraction is more than 1.

d. The picture shows 15 parts shaded. There are 27 parts in each group.

e. $\frac{6}{7} \times \frac{4}{7} =$  f. $\frac{6}{7} - \frac{4}{7} =$  g. $\frac{6}{7} + \frac{4}{7} =$  h. $7 = \frac{\quad}{1} = \frac{\quad}{9} = \frac{\quad}{4}$

## Part 9

**Write each fraction as a division problem and work the problem.**

a. $\frac{147}{7}$  b. $\frac{36}{9}$  c. $\frac{1218}{3}$

## Part 10

**Make a number family for each problem and figure out the answer.**

a. There were 43 days that had rain. The rest of the days had no rain. There were 134 days in all. How many days did not have rain?

b. In the forest are fir trees and trees that are not firs. There are 789 fir trees. There are 1456 other trees. How many trees were there in all?

c. Milly collected 348 stamps. Jan collected 113 more stamps than Milly collected. How many stamps did Jan collect?

d. Andy dug up 565 worms. Greg dug up 239 fewer worms than Andy did. How many worms did Greg dig up?

## Part 11

**Copy each problem and work it. Check your work with a calculator.**

a. $\begin{array}{r} 623 \\ \times \quad 4 \\ \hline \end{array}$  b. $\begin{array}{r} 76 \\ -\ 49 \\ \hline \end{array}$  c. $\begin{array}{r} 130 \\ \times\ 75 \\ \hline \end{array}$

# Lesson 64

## Part 1

a. There were 3 girls for every 4 boys.

b. There were 5 cars for every 2 trucks.

c. There were 16 cats for every 7 dogs.

d. There were 12 packs in every 5 cartons.

e. There were 6 flowers for every 9 bugs.

## Part 2

a. 7 x ■ = 56    b. 7 x ■ = 42    c. 3 x ■ = 21

d. 6 x ■ = 30    e. 4 x ■ = 24    f. 3 x ■ = 24

## Part 3

These tables are supposed to show the blue birds and red birds in two places.

a.

| | Forest | City | Total for both places |
|---|---|---|---|
| Blue birds | 70 | | |
| Red birds | | 35 | |
| Total for both birds | | 89 | 200 |

***Fact***: In the forest, there were 29 more blue birds than red birds.

b.

| | Mountains | Valley | Total for both places |
|---|---|---|---|
| Blue birds | 76 | | |
| Red birds | | 107 | 133 |
| Total for both birds | | | |

***Fact***: In the valley, there were 72 fewer blue birds than red birds.

## Independent Work

### Part 4

**Copy each fraction and complete the equation.**

a. $\frac{9}{1}$ b. $\frac{4}{4}$ c. $\frac{36}{4}$ d. $\frac{63}{9}$ e. $\frac{54}{9}$

f. $\frac{10}{2}$ g. $\frac{9}{9}$ h. $\frac{10}{5}$ i. $\frac{10}{1}$ j. $\frac{10}{10}$

### Part 5

**Write the fraction for the first picture. Then write an equation to show the fractions in the row that are equal to the first fraction.**

a.

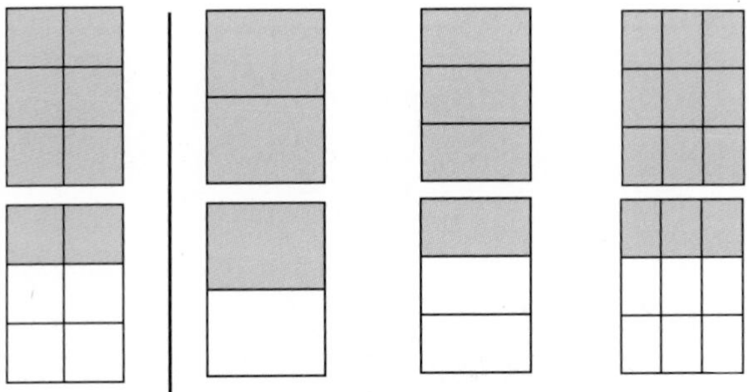

b.

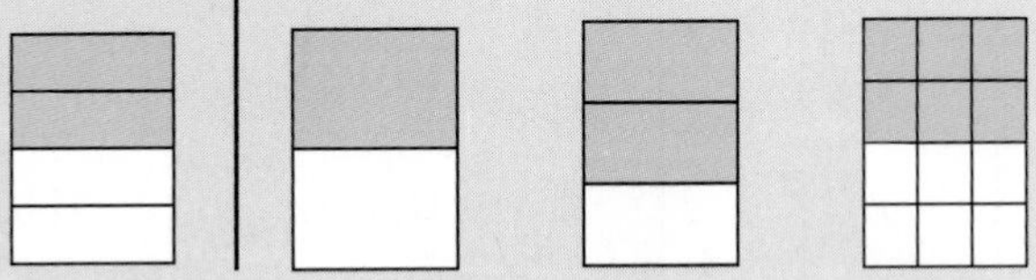

### Part 6

**Write the addition or multiplication fact for each group of numbers. Write each equation so the equal sign comes after the first number.**

a. 3, 15, 12 b. 6, 3, 2 c. 16, 16, 0 d. 8, 24, 3

**Part 7** Copy the number line. Write the fractions for the whole numbers.

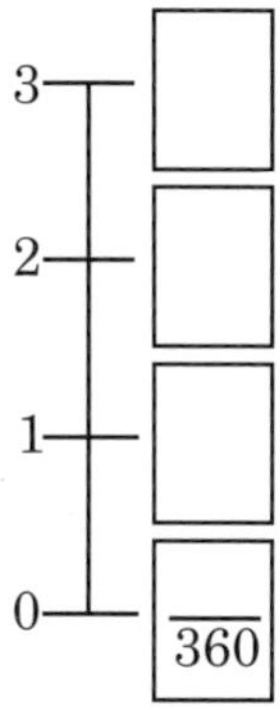

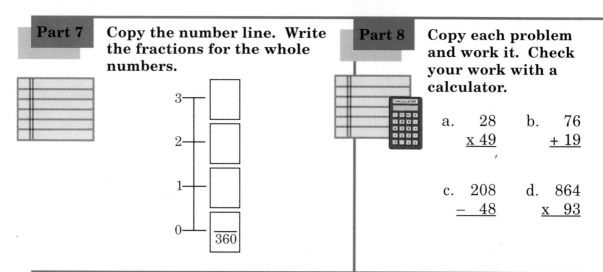

**Part 8** Copy each problem and work it. Check your work with a calculator.

a. 28 x 49

b. 76 + 19

c. 208 − 48

d. 864 x 93

**Part 9** Make a number family for each problem and answer the question.

a. There were 1501 animals. 479 of them were not horses. The rest were horses. How many were horses?

b. Some of the students were outside and the rest were inside. There were 234 outside. There were 167 inside. How many students were there in all?

c. Some of the vehicles were trucks. There were 568 vehicles. 197 of them were not trucks. How many were trucks?

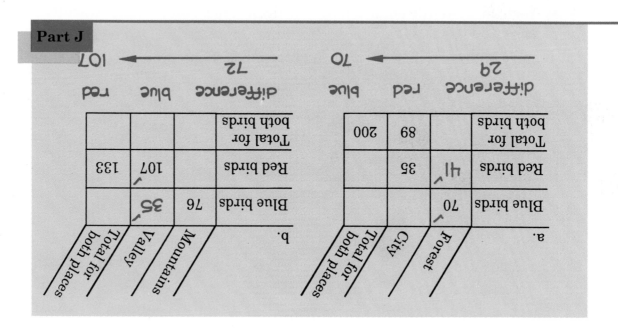

**Part J**

a.

| | Forest | City | Total for both places |
|---|---|---|---|
| Blue birds | 70 | | |
| Red birds | 41 | 35 | |
| Total for both birds | | 89 | 200 |

difference red blue

29 → 70

b.

| | Mountains | Valley | Total for both places |
|---|---|---|---|
| Blue birds | 76 | 35 | |
| Red birds | | 107 | 133 |
| Total for both birds | | | |

difference blue red

72 → 107

# Lesson 65

## Part 1

a. $\frac{38}{9}$ b. $\frac{27}{4}$ c. $\frac{34}{5}$ d. $\frac{23}{4}$ e. $\frac{69}{9}$

## Part 2

- You've learned that if you multiply a fraction by 1, you'll end up with the same value you started with.
- If you don't multiply the first fraction by 1, you don't end up with the same value you started with.
- You've also learned that if two pictures of fractions are equal, they have shaded areas that are the same size.
- Here are pictures of two fractions that are equal:

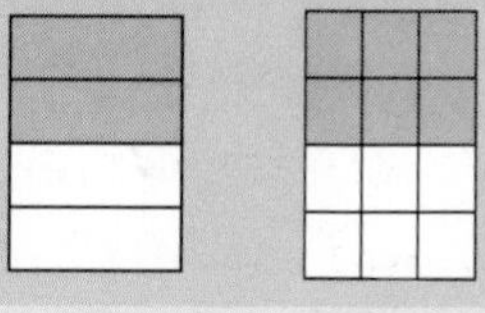

- You can write the equation for these pictures. Your equation will show the two fractions that are equal.
- The equation will also show the fraction of 1 you multiply by.
- Here are the fractions for the two pictures:

$$\frac{2}{4} = \frac{6}{12}$$

- To find the fraction that equals 1, you say the multiplication problem for the bottom and for the top.

$$\frac{2}{4} \times \boxed{\frac{\ }{\ }} = \frac{6}{12}$$

- There are 3 times as many parts in each unit. And there are 3 times as many parts shaded.

$$\frac{2}{4} \times \boxed{\frac{3}{3}} = \frac{6}{12}$$

a.

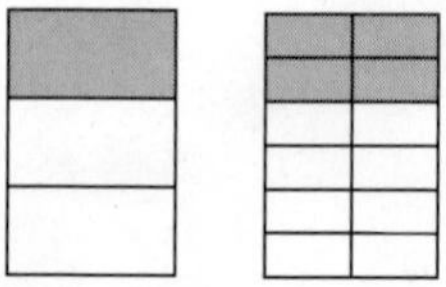

b.

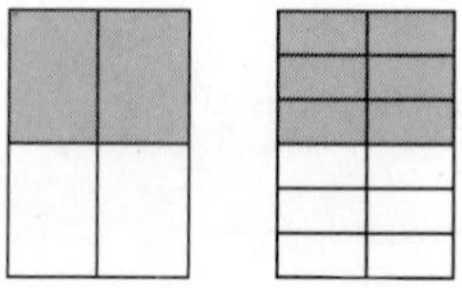

c.

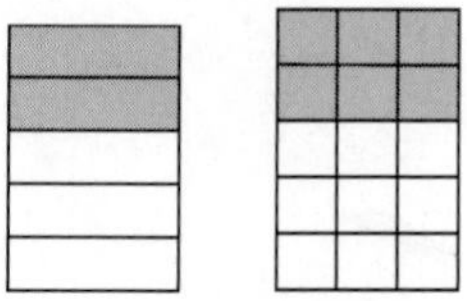

## Part 3

a. 7 leaves fell for every 5 seeds that fell.

b. There are 6 dogs for every 9 cats.

c. In a hospital, there are 3 nurses for every 17 patients.

d. There are 5 cones on every 8 trees.

## Part 4

a. 5 x ■ = 355  b. 6 x ■ = 360  c. 3 x ■ = 618  d. 4 x ■ = 840

## Independent Work

## Part 5

**For each problem, make a number family and answer the question.**

a. Joe had $15.00. He bought one of the items. He left the store with $12.63. Which item did he buy?

b. Jan needs $1.25 more to buy the scissors. How much money does she have?

c. Jane has $34.56. If she buys the calendar, how much money will Jane have left?

## Part 6

**Write each problem so the equal sign comes after the first number. Then complete the equation.**

a. 72 – ■ = 70  b. 6 ■ 5 = 30  c. 30 ■ 20 = 10

## Part J

a. 5|355  b. 6|360  c. 3|618  d. 4|840

# Lesson 66

## Part 1

a. $\frac{33}{4}$ b. $\frac{33}{5}$ c. $\frac{15}{3}$ d. $\frac{15}{4}$ e. $\frac{8}{2}$

## Part 2

a.

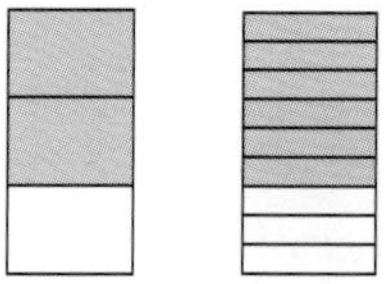

b.

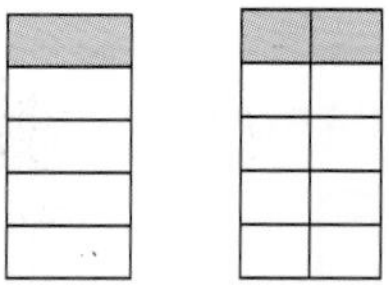

## Part 3

| | |
|---|---|
| 6 | 12 |
| | 24 |
| | 36 |
| | 48 |
| | 60 |

## Independent Work

## Part 4

**Use the table in part 2 of your workbook to answer these questions.**

a. Which restaurant sold 214 tacos?

b. How many tacos and hot dogs did Joe's Stand sell?

c. Were more tacos sold at both restaurants or more hot dogs sold at both restaurants?

## Part 5

**Write the complete equations with the boxed number below.**

a. 39 = 9 x ■ + ■ □

b. 64 = 7 x ■ + ■ □

c. 22 = 4 x ■ + ■ □

d. 22 = 3 x ■ + ■ □

## Part 6 Write the names and the fraction for each sentence.

a. There are 7 riders for every 11 horses.

b. There are 12 pens for every 5 students.

c. Every 3 cases holds 9 bottles.

## Part 7 Copy each problem and work it. Check your work with a calculator.

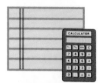

a. $\begin{array}{r} 376 \\ +\ 109 \\ \hline \end{array}$

b. $\begin{array}{r} 208 \\ -\ 118 \\ \hline \end{array}$

c. $\begin{array}{r} 64 \\ \times\ 93 \\ \hline \end{array}$

## Part 8 Make a number family for each problem and figure out the answer.

a. A truck traveled 560 fewer miles than a boat traveled. The boat traveled 990 miles. How many miles did the truck travel?

b. Mr. Jones drove in the morning and in the afternoon. He drove 608 miles in all. If he drove 345 miles in the morning, how many miles did he drive in the afternoon?

## Part 9 Figure the area of each rectangle.

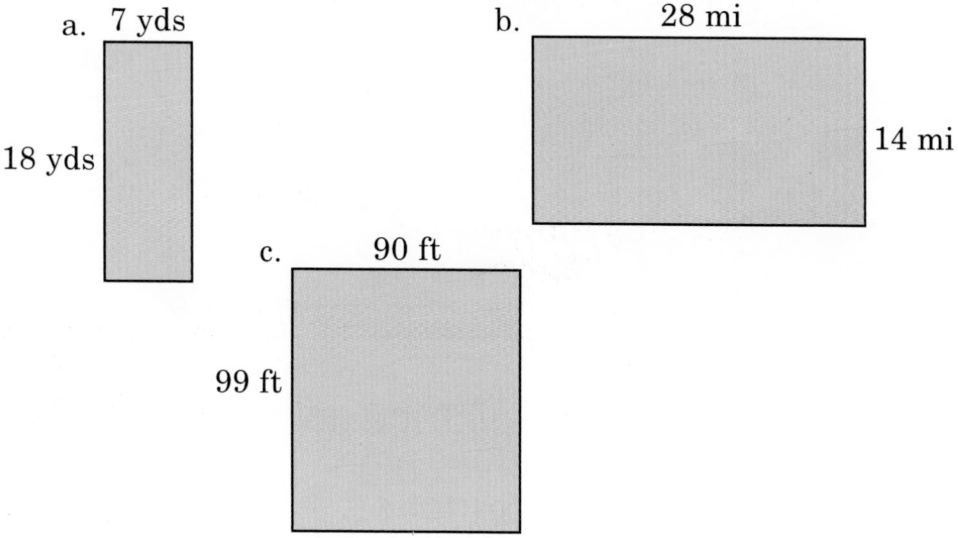

## Part 10 Write each fraction as a division problem and figure out the answer.

a. $\frac{505}{5}$ b. $\frac{357}{7}$ c. $\frac{216}{2}$

## Part J

| | | |
|---|---|---|
| a. | $4 \times 7 = 28$ | $4\overline{)28}$ = 7 |
| b. | $3 \times 8 = 24$ | $3\overline{)24}$ = 8 |
| c. | $2 \times \underline{8} = 16$ | $2\overline{)16}$ = 8 |
| d. | $4 \times 2 = 8$ | $4\overline{)8}$ = 2 |

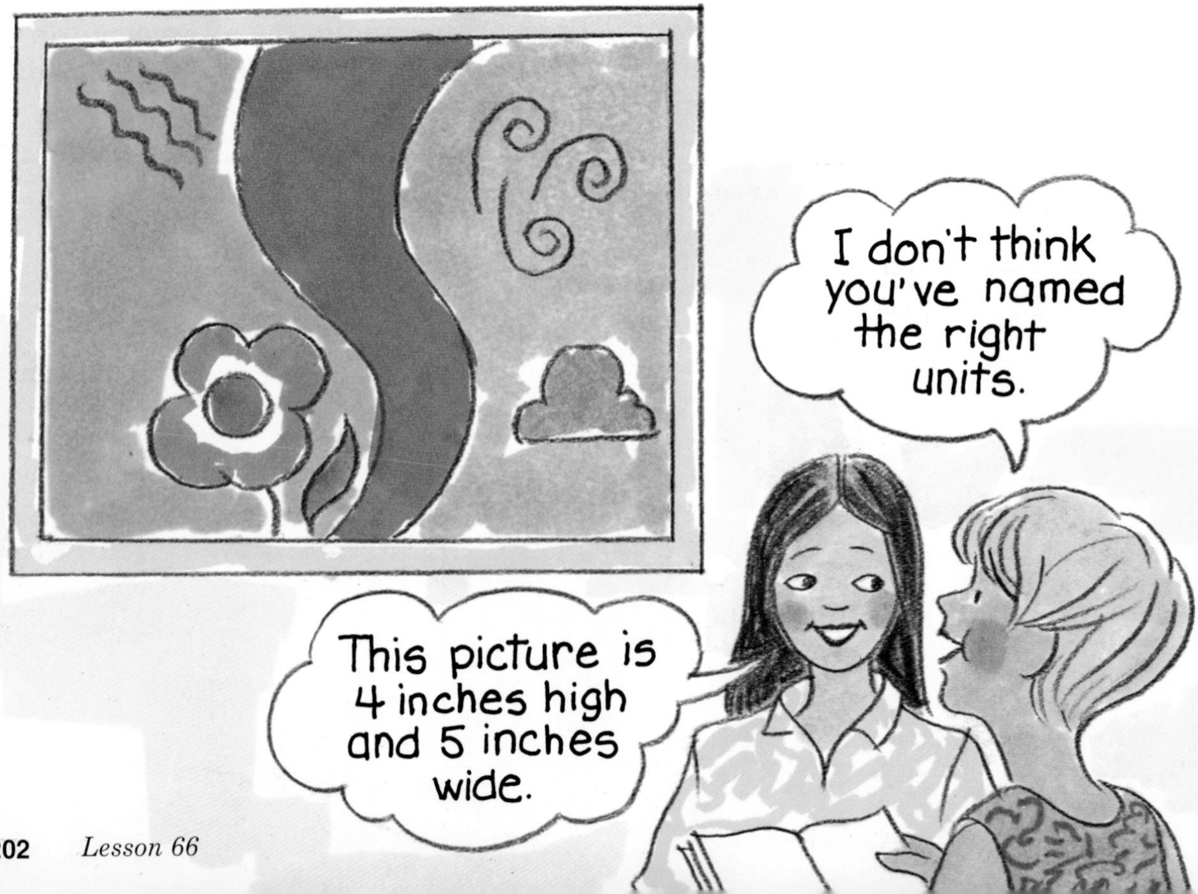
I don't think you've named the right units.
This picture is 4 inches high and 5 inches wide.

# Lesson 67

## Part 1

a. $\frac{36}{4}$ b. $\frac{36}{5}$ c. $\frac{36}{9}$ d. $\frac{13}{5}$ e. $\frac{13}{2}$

## Part 2

This table is supposed to show the number of grapes that were ripe and the number of grapes that were not ripe in Brandon Vineyard and Eden Vineyard.

| | Ripe | Not ripe | Total grapes |
|---|---|---|---|
| Brandon | | | |
| Eden | | | |
| Total for both vineyards | | | |

***Questions***

a. Which vineyard had fewer ripe grapes?

b. How many total grapes were there in both vineyards?

c. Which is greater, the total of ripe grapes or the total of not-ripe grapes?

***Facts***

1. In Brandon Vineyard, 1041 grapes are not ripe.
2. In Eden Vineyard, 7084 grapes are ripe.
3. In Brandon Vineyard, there are 9500 ripe grapes.
4. In Eden Vineyard, there are 2579 more ripe grapes than not-ripe grapes.

## Part 3

a. 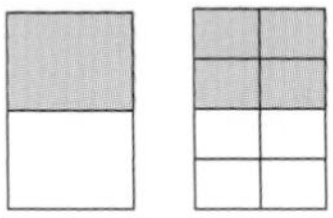

b. 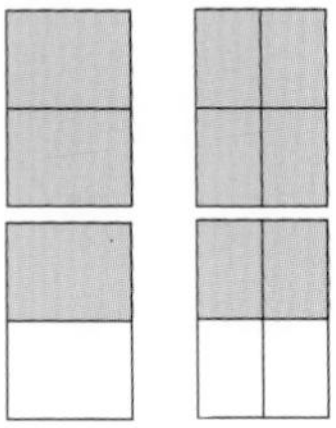

## Part 4

- If a number family shows a fraction and a whole number, you have to change the whole number into a fraction.
- The bottom number of the fraction you change the whole number into is the same as the bottom number of the other fraction in the family.
- Here's a number family with a fraction and a whole number: $\frac{3}{7} \longrightarrow 1$
- The whole number has to be changed into a fraction with 7 as the bottom number.
- Here's the family with 1 crossed out and $\frac{7}{7}$ written in its place. $\frac{3}{7} \longrightarrow \cancel{1}\ \frac{7}{7}$
- Now you just work the problem on top to find the missing number. The missing fraction is $\frac{4}{7}$. $\frac{3}{7} \xrightarrow{\frac{4}{7}} \cancel{1}\ \frac{7}{7}$
- Remember, work the problem just like other problems that have mixed numbers. You change the whole number into a fraction.

## Independent Work

## Part 5

**Copy each problem and work it.**

a. $\frac{3}{5} \times \frac{7}{8} = \blacksquare$  b. $\frac{1}{7} + \frac{8}{7} = \blacksquare$  c. $\frac{3}{3} - 1 = \blacksquare$  d. $\frac{2}{5} \times \frac{1}{5} = \blacksquare$

## Part 6

**Copy the table. For each row, write a division problem and an answer. Write a fraction and the whole number it equals.**

| | |
|---|---|
| $\overline{)\quad}$ | $\frac{\square}{\square} = \square$ |
| | $\frac{18}{2} =$ |
| $7\overline{)63}$ | |
| | $\frac{24}{4} =$ |
| | $\frac{12}{3} =$ |

## Part 7

**Write the names and the fraction for each sentence.**

a. For every 8 fish there were 2 dolphins.

b. A restaurant served 6 pops for every 10 hamburgers.

c. Joe mowed his lawn 3 times in every 4 weeks.

## Part 8

**Copy each problem and write the answer.**

a. 9 x 7 = ■ b. 7 x 6 = ■ c. 9 x 6 = ■ d. 9 x 8 = ■

e. 3 x 8 = ■ f. 8 x 5 = ■ g. 7 x 4 = ■ h. 4 x 8 = ■

---

i. $9\overline{)27}$ j. $9\overline{)54}$ k. $7\overline{)28}$ l. $7\overline{)42}$ m. $4\overline{)28}$

n. $8\overline{)40}$ o. $9\overline{)36}$ p. $7\overline{)21}$ q. $7\overline{)63}$

## Part 9

**Copy the problems and work them.**

a. $4\overline{)368}$ b. $5\overline{)400}$ c. $2\overline{)216}$

# Lesson 68

## Part 1

a. 6 x 8 = ■  b. 6 x 3 = ■  c. 6 x 6 = ■  d. 6 x 4 = ■

e. 6 x 9 = ■  f. 6 x 2 = ■  g. 6 x 7 = ■  h. 6 x 1 = ■

## Part 2

There are 5 cans in every box.

a. There are 6 fish in each pond.

b. There were 13 frogs on each log.

c. There are 20 books for every student.

## Part 3

a. 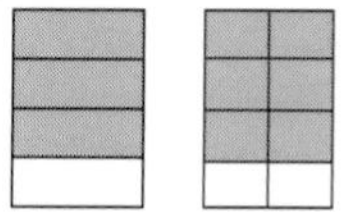

b. 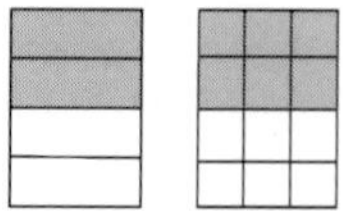

## Part 4

- This division problem is different from those you have worked. 392 ÷ 4. The underlined part is 39. But 39 is not a number for fours.

  4 | 3 9 2 (39 underlined)

- You work the problem by changing the line under 39 into a division sign and writing the number for fours under the sign.

  4 | 3 9 2<br>  | 3 6

- You work the problem under the new division sign and write the answer above. The answer is 9.

  9<br>4 | 3 9 2<br>  | 3 6

- The difference between 39 and 36 is 3. That's the remainder. You carry the remainder and write it in front of the next digit.

- Here's the underlining for the last digit of the answer.

  9 8<br>4 | 3 9 ³2 (32 underlined)<br>  | 3 6

## Independent Work

**Part 5** **Copy the table. Use the information that the facts give to write four numbers in the table. Fill in the rest of the table. Then answer the questions.**

This table is supposed to show the number of ducks and geese seen on Clear Lake and Long Lake.

| | Clear Lake | Long Lake | Total for both lakes |
|---|---|---|---|
| Ducks | | | |
| Geese | | | |
| Total for both birds | | | |

***Facts***

1. The total geese seen on both lakes is 545.
2. 256 ducks were seen on Long Lake.
3. 438 geese were seen on Clear Lake.
4. On Clear Lake, there were 351 fewer ducks than geese.

***Questions***

a. Were there more ducks on Clear Lake or on Long Lake?

b. How many total birds were on Long Lake?

c. How many birds were seen on both lakes?

**Part 6** **Write the fraction for each letter.**

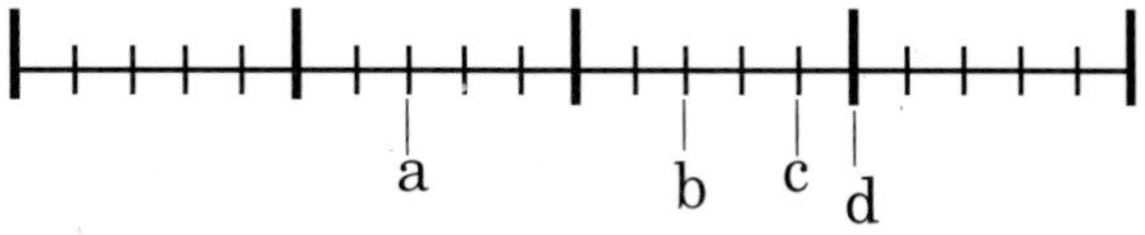


**Part 7** **Copy the mixed number and write the fraction addition equation.**

a. $4 + \frac{2}{9}$ b. $5 + \frac{3}{7}$ c. $7 + \frac{2}{5}$ d. $9 + \frac{3}{4}$

## Part 8

**Copy the table. For each row, write a complete multiplication equation and write the division problem and the answer.**

| | |
|---|---|
| □ x __ = □ | $\overline{)\phantom{00}}$ |
| □ x __ = □ | $3\overline{)27}$ |
| 4 x __ = 28 | $\overline{)\phantom{00}}$ |
| □ x __ = □ | $9\overline{)36}$ |
| □ x __ = □ | $5\overline{)35}$ |

## Part 9

**For each fraction, write the division problem and the answer.**

a. $\frac{248}{2}$  b. $\frac{820}{4}$  c. $\frac{459}{9}$

## Part 10

**Copy each problem and work it. Check your work with a calculator.**

a. $36 \times 74$

b. $57 + 65$

c. $259 \times 68$

d. $765 - 67$

e. $234 + 432$

f. $586 - 507$

g. $432 - 34$

# Lesson 69

## Part 1

**Each box holds 31 nails.**

a. Every cat has 19 fleas.

b. Every shirt has 8 buttons.

c. There are 8 buttons on every shirt.

d. There are 15 fish in each can.

e. Each baby has 12 teeth.

## Part 2

a. 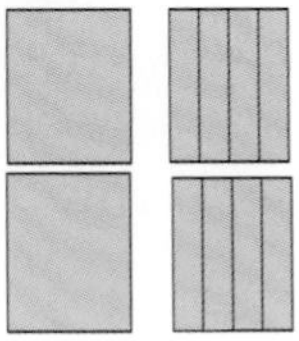

b. 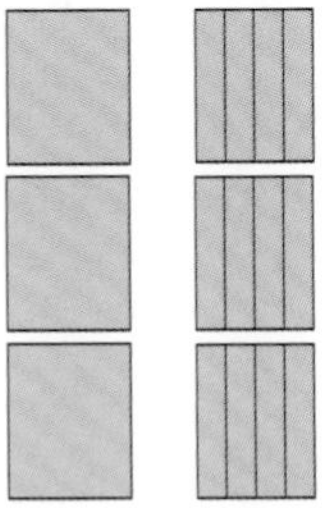

## Part 3

| | | |
|---|---|---|
| | ☐ x ___ = ☐ | $\overline{)\phantom{000}}$ |
| a. | ☐ x ___ = ☐ | $5\overline{)355}$ |
| b. | 4 x ___ = 828 | $\overline{)\phantom{000}}$ |
| c. | 3 x ___ = 396 | $\overline{)\phantom{000}}$ |

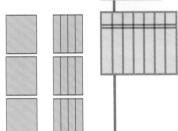

## Independent Work

## Part 4

**Make a number family for each problem and figure out the answer.**

a. Jill read 21 books, and Dale read some other books. Dale read 17 books. How many books did they read?

b. A store sold lamps and tables. The store sold 345 lamps. The store sold 913 pieces of furniture in all. How many tables did the store sell?

## Part 5 Copy and work each problem.

a. $\begin{array}{r} 36 \\ \underline{\times\ 12} \end{array}$

b. $\begin{array}{r} 391 \\ \underline{\times\ \ \ 0} \end{array}$

c. $\begin{array}{r} 486 \\ \underline{\times\ \ \ 1} \end{array}$

d. $\begin{array}{r} 428 \\ \underline{+\ \ \ 8} \end{array}$

## Part 6 Write the addition problem and the answer for each item.

a. A doctor's office bought a stapler, a note pad and a telephone. How much did the doctor's office spend?

b. Jill purchased a radio and a ruler. How much money did she spend?

c. How much did Don spend if he bought a note pad, a ruler and a stapler?

## Part 7 Copy and complete each equation.

a. $\frac{1}{3} \times \frac{3}{7} =$

b. $\frac{12}{5} - \frac{12}{5} =$

c. $1 - \frac{2}{9} =$

d. $\frac{180}{3} \times \frac{1}{10} =$

e. $\frac{17}{9} - 1 =$

## Part 8 Write the answer to each problem.

a. 6 x 8 = ■

b. 6 x 4 = ■

c. 6 x 6 = ■

d. 6 x 7 = ■

e. 6 x 9 = ■

## Part 9

**Copy each problem and write the complete equation.**

a. $4 = \frac{\quad}{7} = \frac{\quad}{6} = \frac{\quad}{9} = \frac{\quad}{1} = \frac{\quad}{4}$

b. $7 = \frac{\quad}{5} = \frac{\quad}{9} = \frac{\quad}{8} = \frac{\quad}{3}$

c. $\frac{243}{3} =$

d. $\frac{175}{5} =$

e. $\frac{18}{3} =$

f. $\frac{56}{7} =$

## Part 10

**Copy the table. Use the facts to write four numbers in the table. Then fill in the table.**

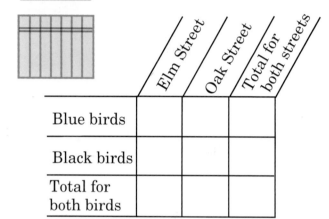

| | Elm Street | Oak Street | Total for both streets |
|---|---|---|---|
| Blue birds | | | |
| Black birds | | | |
| Total for both birds | | | |

***Facts***

1. There was a total of 172 blue birds.
2. There were 450 black birds on Oak Street.
3. The total number of black birds was 584 more than the total number of blue birds.
4. There were 51 blue birds on Elm Street.

## Part 11

**Copy the mixed number and write the fraction addition equation.**

a. $24 + \frac{2}{3}$ b. $5 + \frac{4}{5}$ c. $7 + \frac{1}{4}$ d. $4 + \frac{7}{8}$

## Part J

a. $\frac{1}{2}$ $\frac{7}{2}$ → ~~4~~ $\frac{8}{2}$

b. $\frac{2}{5}$ ~~3~~ $\frac{15}{5}$ → $\frac{17}{5}$

c. $\frac{12}{14}$ $\frac{2}{14}$ → ~~1~~ $\frac{14}{14}$

d. ~~1~~ $\frac{31}{31}$ $\frac{9}{31}$ → $\frac{40}{31}$

e. $\frac{12}{3}$ $\frac{6}{3}$ → ~~6~~ $\frac{18}{3}$

# Test 7

## Part 5

a. 3 x ■ = 219 b. 2 x ■ = 64 c. 4 x ■ = 412

## Part 6

a. $\frac{4}{7} \xrightarrow{\quad 1 \quad}$ b. $\xrightarrow{\quad \frac{2}{5} \quad} 6$ c. $1 \xrightarrow{\qquad} \frac{26}{4}$

## Part 7

a. $9\overline{)360}$ b. $3\overline{)615}$ c. $4\overline{)48}$

## Part 8

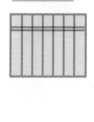

a. June has $10.75. She wants to buy the sandals. How much more money does she need?

b. The hat costs $8.10 less than one of the items. Which item?

c. Sam has $14.83 more than the cost of the jacket. How much does Sam have?

# Lesson 70

## Part 1

a.

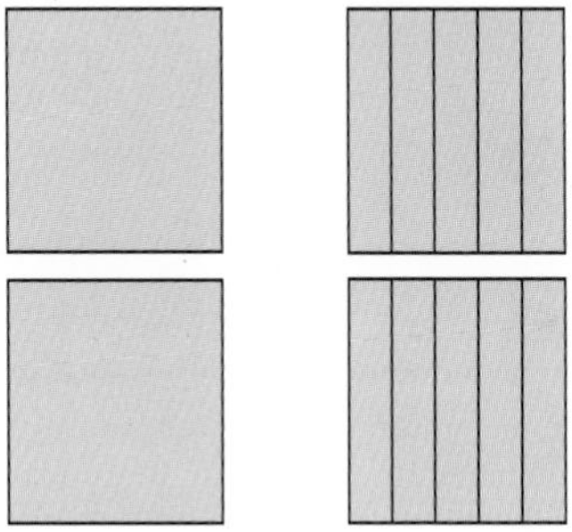

b.

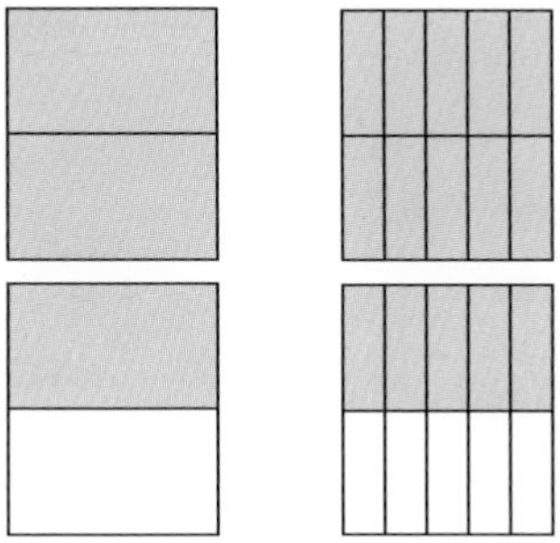

## Part 2

a. $4 + \frac{3}{5}$ b. $7 + \frac{1}{2}$ c. $4 + \frac{6}{8}$ d. $16 + \frac{4}{7}$

No, Junior, that's not how you make mixed numbers.

# Lesson 71

## Part 1

a. $\frac{1}{7} \longrightarrow 3$

b. $1 \quad \frac{7}{15} \longrightarrow$

c. $\xrightarrow{\quad \frac{2}{9}\quad} 1$

d. $\frac{11}{12} \quad 2 \longrightarrow$

## Part 2

a. 3 x ■ = 624

b. 4 x ■ = 880

c. 7 x ■ = 497

## Part 3

- Here's an important rule about multiplication: If you multiply by **more than 1,** you end up with **more than you start out with,** and if you multiply by **less than 1,** you end up with **less than you start out with.**
- Here's a problem:
  You can't see the number that 3 is multiplied by. But you can figure out whether it is more than 1 or less than 1.

  3 x ■ = 2
- You just compare the number you start out with and the number you end up with. The number you start with is 3. The number you end up with is 2.
- The number you end up with is less than the number you start with, so the number you multiply by is less than 1.

  3 x ■ = 2
  **less than 1**
- Here's another problem:
  You ask: Do you end up with more than you start with or less than you start with?

  3 x ■ = 4
- If you end up with **more** than you start with, you're **multiplying by more than 1.** If you end up with **less** than you start with, you're **multiplying by less than 1.**

## Independent Work

### Part 4

**Make a number family for each problem and answer the question.**

a. In a garden, there were red roses and 185 roses that were not red. There were 520 roses in all. How many red roses were there?

b. Joe worked in a grove of trees. Joe chopped down 27 trees. There were 92 trees still standing. How many trees were there in all?

### Part 5

**Figure out the area for each rectangle.**

a.
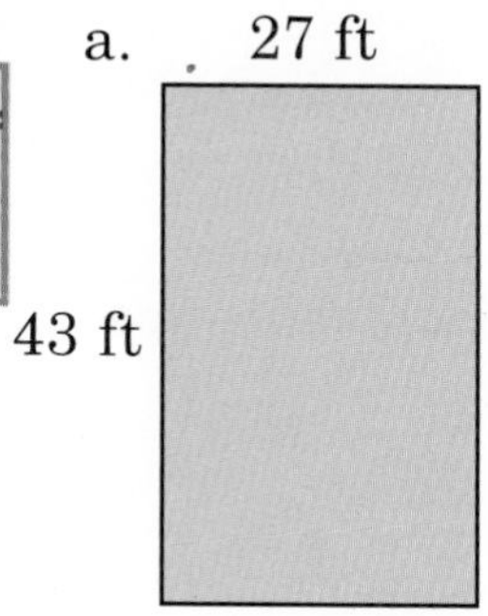


b.
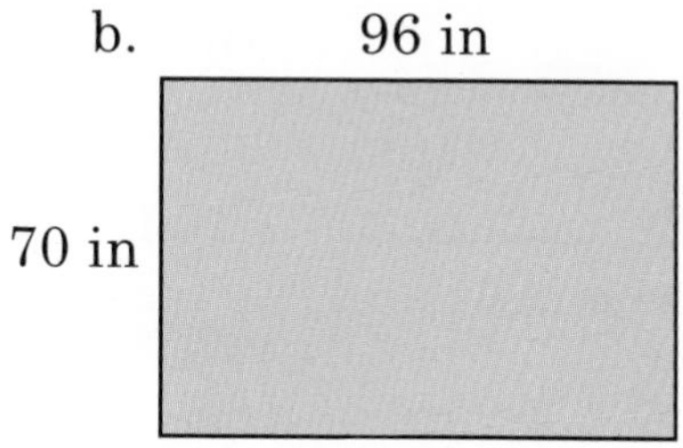


c.
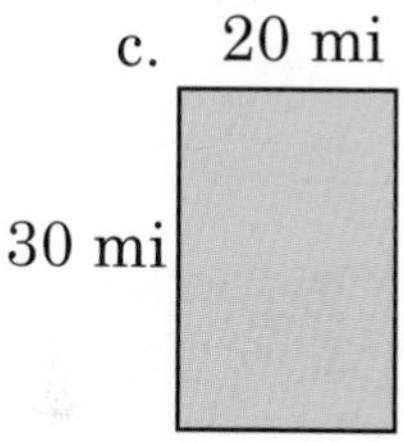


### Part 6

**For each row, write a division problem and the answer. Write the fraction and the whole number it equals.**

| | $\overline{)\phantom{0}}$ | $\frac{\square}{\square} = \square$ |
|---|---|---|
| a. | $3\overline{)96}$ | $\frac{\phantom{0}}{\phantom{0}} = \square$ |
| b. | $5\overline{)250}$ | $\frac{\phantom{0}}{\phantom{0}} = \square$ |
| c. | | $\frac{612}{2} = \square$ |

## Part 7

**Copy each equation and complete the fractions.**

a. $7 = \frac{\quad}{4} = \frac{\quad}{7} = \frac{\quad}{9} = \frac{\quad}{8}$

b. $6 = \frac{\quad}{5} = \frac{\quad}{1} = \frac{\quad}{9} = \frac{\quad}{4}$

## Part 8

**Copy each mixed number. Below, write the equation that shows the fraction each mixed number equals.**

a. $32 + \frac{5}{6}$

b. $7 + \frac{1}{24}$

## Part 9

**Copy each fraction and complete the equation with a mixed number.**

a. $\frac{75}{9} =$

b. $\frac{11}{9} =$

c. $\frac{22}{3} =$

## Part 10

**Write the names and the fraction for each sentence.**

a. Each case contains 48 bottles.

b. There were 4 horses in every stall.

c. There were 6 vowels for every 10 letters.

d. Every goat ate 11 carrots.

# Lesson 72

## Part 1

a. $2\overline{)706}$ b. $4\overline{)460}$ c. $3\overline{)1806}$ d. $3\overline{)1410}$

## Part 2

a. 7 x ■ = 420 b. 5 x ■ = 455 c. 3 x ■ = 639

## Independent Work

## Part 3

**For each row, write a division problem and the fraction equation.**

| | $\overline{)\phantom{0}}$ | $\frac{\square}{\square} = \square$ |
|---|---|---|
| a. | $7\overline{)35}$ | |
| b. | | $\frac{820}{4} =$ |
| c. | $5\overline{)155}$ | |
| d. | $9\overline{)270}$ | |

## Part 4

**Copy each family with the whole number and fraction. Cross out the whole number and write the fraction it equals. Then figure out the missing fraction and box it.**

a. $5 \xrightarrow{\quad \_\_ \quad} \frac{20}{3}$ b. $\frac{7}{16} \xrightarrow{\quad \_\_ \quad} 1$ c. $\frac{24}{4} \xrightarrow{\quad 2 \quad} \_\_$

## Part 5

Copy each mixed number. Below, write the complete equation.

a. $2 + \frac{1}{8}$   b. $3 + \frac{5}{36}$

## Part 6

Write the complete equation to show the mixed number.

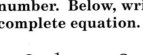

a. $\frac{11}{3} =$   b. $\frac{30}{28} =$   c. $\frac{59}{7} =$

## Part 7

Copy each problem. Write the number 7 is multiplied by and the remainder to complete each equation.

a. 11 = 7 x ■ + R ■ (□)   b. 27 = 7 x ■ + R ■ (□)

c. 4 = 7 x ■ + R ■ (□)   d. 45 = 7 x ■ + R ■ (□)

## Part 8

Write the answer to each problem.

a. 6 x 4 = ■   b. 8 x 6 = ■   c. 8 x 8 = ■   d. 9 x 7 = ■

e. 6 x 7 = ■   f. 6 x 3 = ■   g. 6 x 6 = ■

## Part 9

Copy each item and complete the equation.

a. $12 = \frac{\quad}{3} = \frac{\quad}{1} = \frac{\quad}{7}$   b. $24 = \frac{\quad}{2} = \frac{\quad}{5} = \frac{\quad}{3}$

# Lesson 73

## Part 1

a. $4 \times ■ = 812$    b. $3 \times ■ = 936$    c. $9 \times ■ = 540$

## Part 2

a.

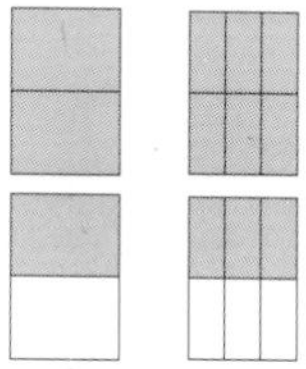

b.

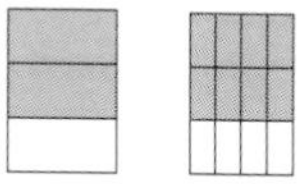

## Part 3

**Sample Problem**

$$\frac{3}{4} = \frac{624}{832}$$

a. $\frac{2}{5} = \frac{122}{305}$    b. $\frac{3}{7} = \frac{63}{147}$    c. $\frac{9}{2} = \frac{936}{208}$

## Part 4

The $x$ arrow is along the bottom. The $y$ arrow is up.

To find the point, you first go along the $x$ arrow. Then up for $y$.

a. $(x = 6,\ y = 10)$

b. $(x = 4,\ y = 9)$

c. $(x = 2,\ y = 4)$

d. $(x = 9,\ y = 5)$

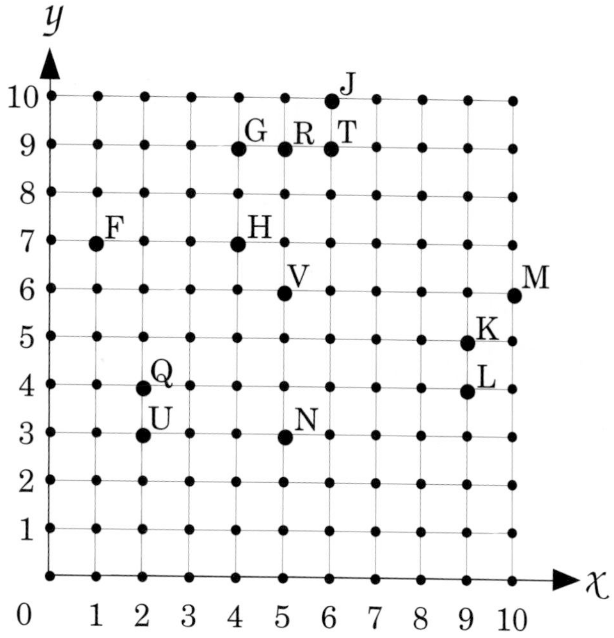

## Part 5

a. There are 5 birds for every 3 cats. How many cats are there if there are 25 birds?

b. There are 8 geese for every 5 ducks. How many ducks are there if there are 32 geese?

c. There are 9 snakes in every garden. There are 8 gardens. How many snakes are there?

## Independent Work

## Part 6

**Copy and complete each number family with three fractions. Box the answer.**

a. $\frac{13}{15} \xrightarrow{\quad \underline{\quad} \quad} 1$

b. $\frac{3}{10} \xrightarrow{\quad 4 \quad} \underline{\quad}$

## Part 7

**Copy each equation and complete it.**

a. 18 = 20 ■ 2  b. 5 ■ 9 = 45  c. 81 = 9 ■ 9  d. $\frac{47}{7} =$  e. $\frac{120}{6} =$

f. $12 = \frac{\quad}{1} = \frac{\quad}{3} = \frac{\quad}{15}$

g. $6 = \frac{\quad}{9} = \frac{\quad}{5} = \frac{\quad}{8} = \frac{\quad}{7}$

## Part 8

**Copy each problem and write the complete equation.**

a. $\frac{12}{5} - 1 =$  b. $\frac{8}{3} \times 6 =$  c. $\frac{11}{4} - \frac{5}{4} =$

d. $1 \times \frac{5}{7} =$  e. $\frac{8}{10} \times \frac{3}{2} =$  f. $\frac{16}{4} - 4 =$

## Part 9

**For each problem, draw the diagram. Figure out the area.**

a. A fence is 11 feet high and 290 feet long. What is the area of the fence?

b. The side of a building is 34 meters high and 78 meters wide. What is the area of the side?

# Lesson 74

## Part 1

- You've learned to work with fractions.
- The bottom number of a fraction has a special name–**denominator.** **denominator** → $\frac{8}{4}$
- The demominator of $\frac{8}{4}$ is **4.**
- The denominator is the number you divide by. $4\overline{)8}$

a. $2 + \frac{8}{10}$

b. $7 + \frac{2}{9}$

c. $8 + \frac{1}{4}$

d. $3 + \frac{0}{5}$

## Part 2

a. There are 7 balls for every 10 children. If there are 50 children, how many balls are there?

b. There are 9 dogs in every kennel. How many kennels are there if there are 27 dogs?

c. There are 9 sides for every 2 figures. If there are 63 sides, how many figures are there?

You can make figures with 3 sides each.
But I don't know how many figures these sides will make.

## Part 3

- Some number family problems tell about getting more and then getting less.
- Here's the number family you make for those problems:

  **end up** **out** **in** →

- The first thing that happens is shown by the big number. That's the number that goes **in.**
- The next thing that happens is shown by the number for **out.**
- After that number goes out, there's still some left. That's the number you **end up** with. It's the difference between the number for **in** and the number for **out.**
- Here's a problem:

  A tank is empty. Then 178 gallons flow into the tank. 123 gallons flow out of the tank. How many gallons are still in the tank?

- Here's the family with the numbers:

  **end up** **out** **in**
  **123** → **178**

- To find the number for **end up** with, you figure out the missing number. The answer is 55 gallons.

$$\begin{array}{r} 178 \\ -123 \\ \hline 55 \end{array}$$ **gallons**

- Remember, something goes in; something goes out; and you end up with some. You can use a number family with the names **in, out** and **end up** to work any problem of this type.

a. A pocket is empty. $342 goes into that pocket. Then some money goes out of that pocket. The pocket ends up with $58. How much went out?

b. A woman starts without any roses. She picks 78 roses. She sells 45 roses. How many roses does she end up with?

c. Jill had an empty basket. She put some apples in the basket. Then she gave away 37 of those apples. The basket still had 14 apples in it. How many apples did Jill put in the basket?

## Part 4

The $x$ arrow is along the bottom. The $y$ arrow is up.

To find the point, you first go along the $x$ arrow. Then up for $y$.

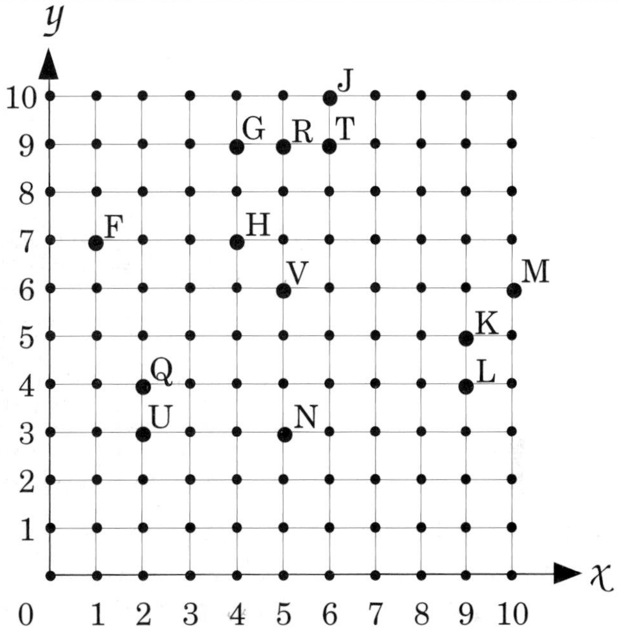


a. $(x = 5, y = 6)$

b. $(x = 4, y = 7)$

c. $(x = 9, y = 4)$

## Independent Work

## Part 5

**Copy each problem. Complete the equation to show the fraction of 1 and the equivalent fraction.**

a. $\frac{1}{3} \times \blacksquare = \frac{\blacksquare}{12}$

b. $\frac{9}{4} \times \blacksquare = \frac{\blacksquare}{28}$

c. $\frac{2}{7} \times \blacksquare = \frac{84}{\blacksquare}$

d. $\frac{3}{5} \times \blacksquare = \frac{216}{\blacksquare}$

## Part 6

**Copy each fraction. Complete the equation to show the mixed number each fraction equals.**

a. $\frac{10}{4} =$

b. $\frac{60}{9} =$

c. $\frac{8}{7} =$

d. $\frac{21}{4} =$

## Part 7

**Copy each problem and write the answer.**

a. 8 x 9 = ■
b. 6 x 9 = ■
c. 6 x 6 = ■
d. 8 x 8 = ■
e. 9 x 9 = ■

f. 7 x 9 = ■
g. 7 x 8 = ■
h. 7 x 7 = ■
i. 6 x 7 = ■
j. 6 x 8 = ■

k. $7\overline{)63}$
l. $9\overline{)54}$
m. $6\overline{)36}$
n. $7\overline{)56}$

o. $7\overline{)42}$
p. $8\overline{)64}$
q. $9\overline{)81}$
r. $7\overline{)49}$

## Part 8

**Copy each number family. Write the missing fraction.**

a. $\underline{\quad}$ $\frac{36}{55}$ ⟶ 1

b. $\frac{67}{84}$ $\underline{\quad}$ ⟶ 1

c. $\frac{407}{635}$ $\underline{\quad}$ ⟶ 1

## Part 9

**Copy the number line and write the fractions for the whole numbers.**

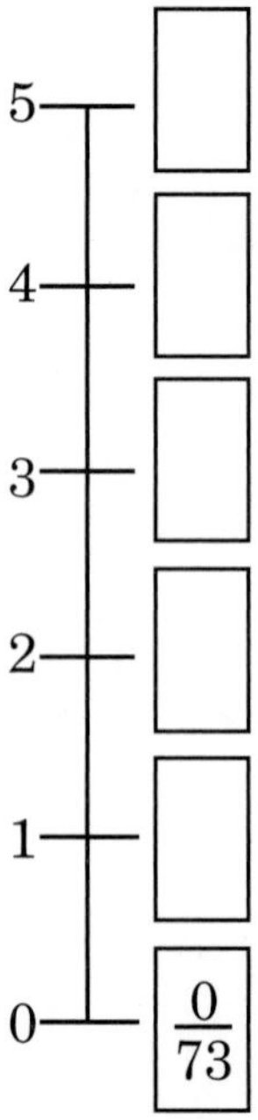

## Part 10

**Figure out the area of each rectangle.**

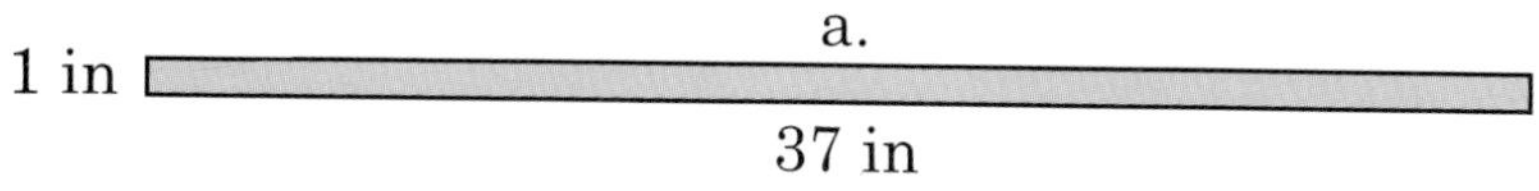

a.
1 in
37 in


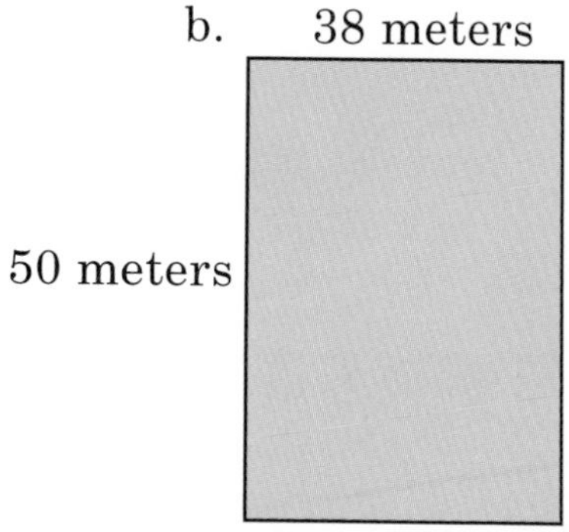

b.
38 meters
50 meters


## Part 11

**Copy each problem. Work the division problem and write the missing number in the multiplication problem.**

a. 3 x ■ = 180

b. 4 x ■ = 840

c. 9 x ■ = 549

d. 7 x ■ = 4949

We can put 9 plants in each carton.
We have 339 plants.
Yeah, but I don't know if we have enough cartons.

# Lesson 75

## Part 1

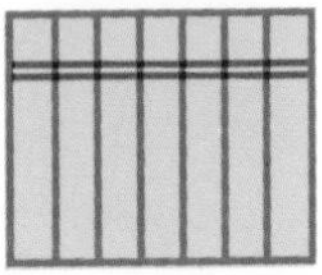

a. $\frac{7}{6} = \frac{\blacksquare}{540}$    b. $\frac{30}{5} = \frac{\blacksquare}{455}$    c. $\frac{4}{9} = \frac{76}{\blacksquare}$

## Part 2

- You've worked problems that tell about the number that goes in, the number that goes out and the number you end up with.
- Sometimes problems tell how much a person had.
- Here's a sentence: A box had 196 stamps in it.
- That sentence tells about the number for **in.** If the box had 196 stamps in it, those stamps must have gone into the box at some time.
- Here are other sentences that tell about the number for **in.**

  Jim had 15.
  A store had 556 pairs of shoes.
  A tank had some oil in it.

- The last sentence doesn't tell you a number for **in,** so you would leave a blank for the big number.

a. A person has \$218. The person spends some money. The person ends up with \$24. How much does the person spend?

b. Jan had some money. She spent \$7.41. She ended up with \$12.50. How much did she have to start with?

c. A store received 137 cases of juice. The store sold 52 cases. How many cases were still in the store?

## Part 3

a. There were 7 perch for every 9 bass. There were 963 bass. How many perch were there?

b. There were 5 frogs for every 6 fish. There were 300 fish. How many frogs were there?

c. A recipe calls for 8 cups of wheat for every 3 cups of rice. How many cups of wheat are needed if there are 78 cups of rice?

## Part 4

- To find the $x$ value for a point, go straight down from the point to the $x$ arrow.
- To find the $y$ value, go straight across from the point to the $y$ arrow.

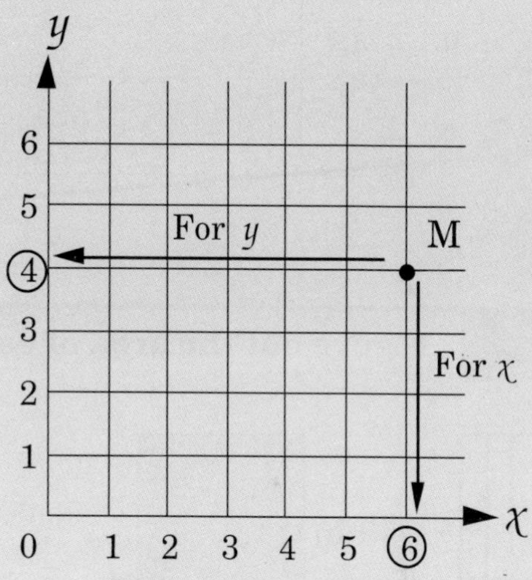


M ($x = 6$, $y = 4$)

## Independent Work

## Part 5

**Copy each problem and write the answer.**

a. 9 x 7 = ■

b. 6 x 8 = ■

c. 7 x 8 = ■

d. 6 x 6 = ■

e. 7 x 6 = ■

f. 6 x 4 = ■

g. 7 x 4 = ■

h. 8 x 8 = ■

i. $7\overline{)28}$

j. $7\overline{)56}$

k. $7\overline{)21}$

l. $6\overline{)36}$

m. $7\overline{)35}$

## Part 6

**For each row, write the division problem and the answer. Then write the multiplication equation.**

| | ⌐ | □ x __ = □ |
|---|---|---|
| a. | $3\overline{)21}$ | |
| b. | $4\overline{)48}$ | |
| c. | | 9 x __ = 927 |

## Part 7

**For each row, write a division problem and the answer. Also write the equation for the fraction and the whole number it equals.**

| | ⌐ | $\frac{□}{□}$ = □ |
|---|---|---|
| a. | $3\overline{)27}$ | |
| b. | $4\overline{)84}$ | |
| c. | $2\overline{)14}$ | |

## Part 8

**Figure out the area of each rectangle.**

a.

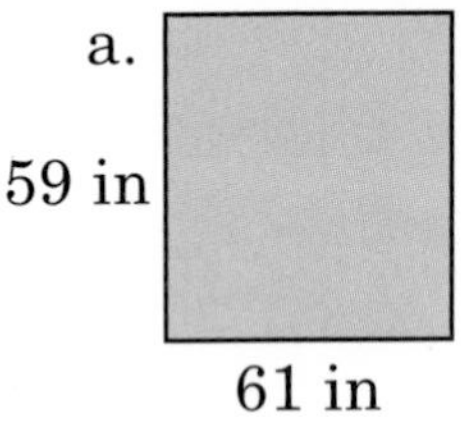


b.

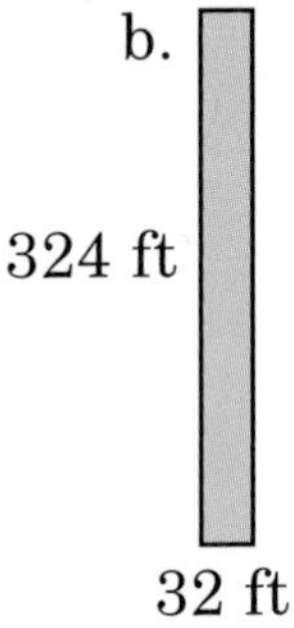


## Part 9

**Make a number family for each problem and figure out the answer.**

a. There were crows and geese in a field. There were 49 crows. There were 219 birds in all. How many geese were in the field?

b. Jill had money in her pocket and money in a jar. She had $7.75 in the jar. She had $19.90 in all. How much money did Jill have in her pocket?

**Part J**

a. $4\overline{)144}$ = 36

b. $4\overline{)6812}$ = 1703

c. $3\overline{)68112}$ = 22704

d. $2\overline{)1058}$ = 529

e. $3\overline{)9315}$ = 3105

# Lesson 76

**Part 1**

a. There were 4 green marbles for every 3 red marbles. How many green marbles were there if there were 24 red marbles?

b. In each pack, there are 6 bottles. If there are 7 packs, how many bottles are there?

c. There are 8 bugs for every 5 frogs. There are 400 bugs. How many frogs are there?

d. There are 2 pounds of water for every 5 pounds of sand. If there are 210 pounds of water, how many pounds of sand are there?

**Part 2**

a. Joe had some money. He spent $86.00 on caps for his team. Joe still had $16.40. How much money did Joe have in the beginning?

b. A farmer had 563 pounds of hay. He fed some hay to his animals. He ended up with 275 pounds of hay. How much hay did he feed to his animals?

c. Sam had $11.20. Then he bought a toy that cost $4.61. How much money did Sam end up with?

**Part 3**

| | |
|---|---|
| 6 | 12 |
| 18 | 24 |
| 30 | 36 |
| 42 | 48 |
| 54 | 60 |

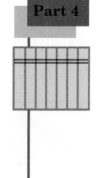

**Part 4**

a. $= \frac{36}{250} \rightarrow 1$

b. $\frac{67}{84} = \rightarrow 1$

c. $\frac{407}{635} = \rightarrow 1$

d. $= \frac{46}{106} \rightarrow 1$

## Independent Work

**Part 5**

**For each row, write the division problem and the answer. Then write the multiplication equation.**

| | □ x __ = □ | $\overline{)\ \ }$ |
|---|---|---|
| a. | | $9\overline{)72}$ |
| b. | | $2\overline{)84}$ |
| c. | 3 x __ = 219 | |

**Part 6**

**For each row, write the division problem and the answer. Also write the fraction and the whole number it equals.**

| | $\overline{)\ \ }$ | $\frac{□}{□}$ = □ |
|---|---|---|
| a. | $7\overline{)63}$ | |
| b. | $7\overline{)49}$ | |
| c. | | $\frac{77}{7}$ = □ |

**Part 7**

**For each problem, make a number family. Use the information in the table to get two numbers in each family. Figure out the missing number and box the answer.**

| | Tall trees | Trees that are not tall | Total trees |
|---|---|---|---|
| River Park | 28 | 84 | 112 |

### *Questions*

a. In River Park, there are 38 fewer squirrels than trees. How many squirrels are in River Park?

b. There are 250 more tall trees in Ocean Park than there are in River Park. How many tall trees are in Ocean Park?

c. In River Park, 18 of the tall trees are redwoods. How many tall trees are not redwoods?

d. Ferns and trees are the only plants in River Park. There were 374 ferns. How many plants are there in all?

**Part 8** Copy the number line and write the fraction for each whole number.

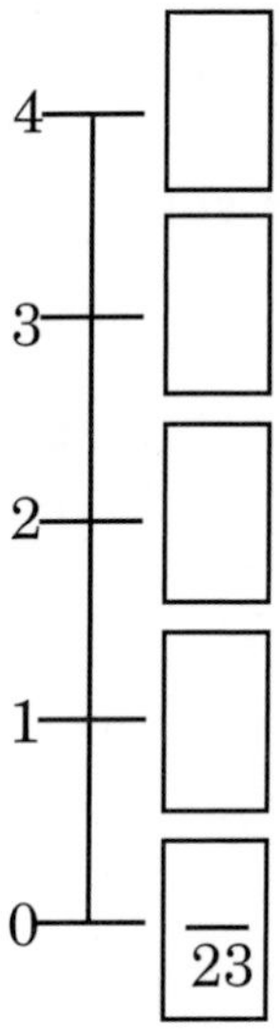


**Part 9** Copy only the problems you can work and complete these equations.

a. $\frac{32}{4} - \frac{4}{5} =$  b. $\frac{16}{9} + \frac{4}{9} =$  c. $\frac{8}{39} - \frac{5}{39} =$

d. $\frac{2}{7} + \frac{2}{5} =$  e. $\frac{16}{9} - \frac{16}{9} =$  f. $\frac{6}{8} + \frac{40}{6} =$

**Part J**

a. $7\overline{)8428}$ = 1204  b. $5\overline{)1030}$ = 206  c. $4\overline{)29292}$ = 7323  d. $5\overline{)5075}$ = 1015

# Lesson 77

## Part 1

a. Some water went into a tank. Then 349 gallons went out of the tank. There were still 663 gallons in the tank. How many gallons did the tank have to begin with?

b. Jan had $45.22. She spent some money. She ended up with $11.11. How much did she spend?

c. A truck started out with 1142 pounds of sand. 1030 pounds of sand were unloaded. How many pounds of sand were still on the truck?

## Part 2

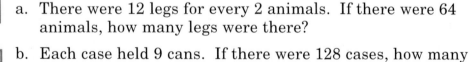

| | |
|---|---|
| 6 | 12 |
| 18 | 24 |
| 30 | 36 |
| 42 | 48 |
| 54 | 60 |

## Part 3

a. There were 12 legs for every 2 animals. If there were 64 animals, how many legs were there?

b. Each case held 9 cans. If there were 128 cases, how many cans were there?

c. There were 9 teeth for every 3 babies. If there were 450 teeth, how many babies were there?

d. Every 5 drawers held 9 pounds. How many pounds were in 540 drawers?

## Part 4

- For some fraction problems, the big number is 1. These problems tell about one whole unit on a number line, or all the parts in one engine, or all the leaves on one tree.
- Here's an example:

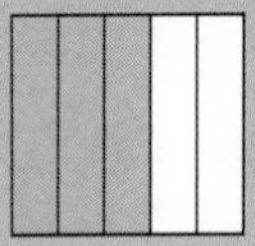

- You can make a fraction number-family for the rectangle. There's one whole rectangle. So the big number is 1.

all
⟶ 1
rectangle

- The small numbers are for the shaded part and the unshaded part.

shaded  not shaded  all
⟶ 1
rectangle

- You write fractions for the shaded parts, the unshaded parts, and the whole rectangle. All the fractions have the same denominator.

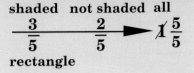

shaded  not shaded  all
$\frac{3}{5}$  $\frac{2}{5}$ ⟶ ~~1~~ $\frac{5}{5}$
rectangle

a.

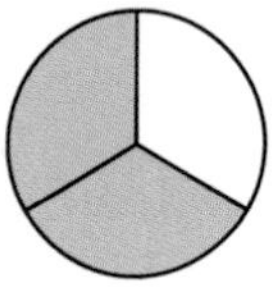

b.

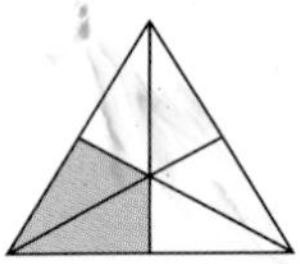

c.

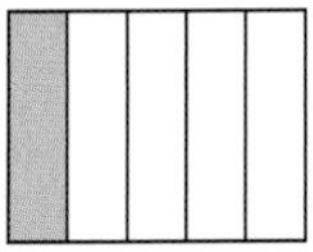

d.

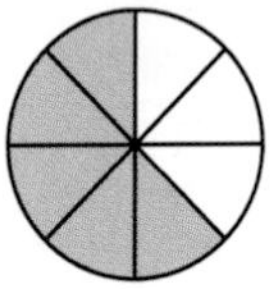

**Part 5**

- Some problems involve letters and **functions**. The letter stands for some number.
- The function is a "rule" that tells you what to do with a number to get an answer.
- Here's a table:
- The letter for this table is $x$. The function is $x + 5$. The function tells you what to do with $x$. You start with $x$ and add 5.

| | Function | Answer |
|---|---|---|
| $x$ | $x + 5$ | ☐ |

- The first column shows different numbers for $x$. If you replace $x$ with 2, the function becomes $2 + 5$.

| | Function | Answer |
|---|---|---|
| $x$ | $x + 5$ | ☐ |
| 2 | $2 + 5$ | |
| 3 | | |
| 7 | | |
| 10 | | |

- If you replace $x$ with 3, the function becomes $3 + 5$.

| | Function | Answer |
|---|---|---|
| $x$ | $x + 5$ | ☐ |
| 2 | $2 + 5$ | |
| 3 | $3 + 5$ | |
| 7 | | |
| 10 | | |

## Independent Work

### Part 6 Figure out the area of each rectangle.

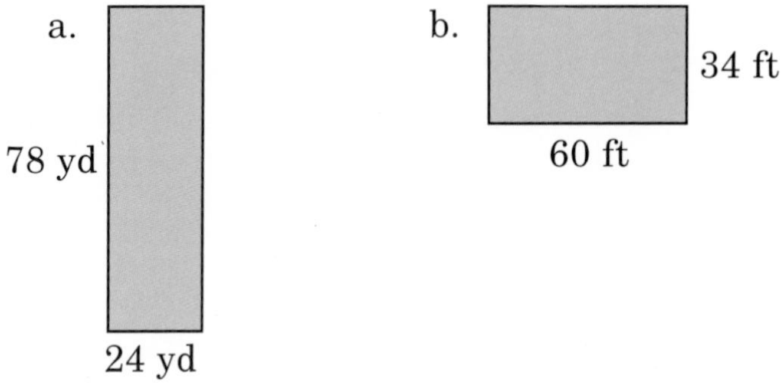


### Part 7 For each problem, make a number family and answer the question.

a. The television costs $44 less than one of the items. Which item?

b. John bought the microwave and had $17 left. How much money did John have to start with?

c. How much more does the stereo cost than the typewriter costs?

### Part 8 Copy each problem and complete the equation.

a. $\frac{72}{9} =$ b. $\frac{}{5} = 10$ c. $\frac{}{3} = 18$ d. $\frac{}{2} = 40$

e. $\frac{270}{3} =$ f. $\frac{3}{5} + 4 =$ g. $\frac{3}{5} \times 4 =$

**Part 9** **Copy each mixed number. Below, write the complete equation.**

a. $4 + \frac{2}{9}$  b. $12 + \frac{3}{4}$  c. $20 + \frac{1}{8}$

**Part 10** **Write the division problem and the answer for each item.**

a. 4 x ■ = 28  b. 9 x ■ = 270  c. 3 x ■ = 615

We can put 9 carrots in each bag.
We have 603 carrots.
If we don't work fast, we won't need very many bags.

# Lesson 78

## Part 1

a. 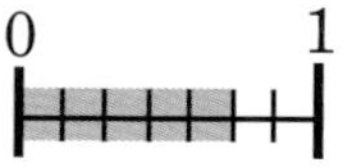

b. 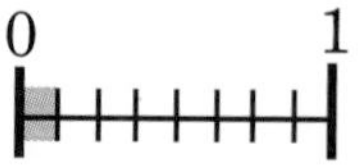

c. 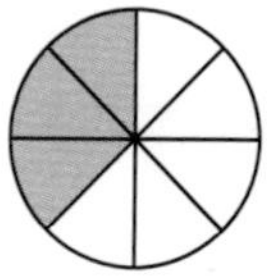

## Part 2

a. ■ + 1 = 70 b. ■ + 1 = 40 c. ■ + 1 = 90

d. ■ + 1 = 600 e. ■ + 1 = 800

## Part 3

a. There are 5 tires for every jeep. There are 101 jeeps. How many tires are there?

b. Each store has the same number of bottles. There are 52 bottles in 3 stores. How many bottles are in 12 stores?

c. There are 9 tons of sand for every 2 trucks. If there are 360 tons of sand, how many trucks are there?

## Part 4

a. Jim had $45.30. He spent $12.95. How much did he end up with?

b. Some people were in a bus. 11 people got out of the bus. The bus still had 29 people in it. How many people were in the bus to start with?

c. There were 85 pencils in a box. Some pencils were removed. There were 38 pencils left in the box. How many pencils had been removed?

## Independent Work

**Part 5** **Copy the number line. Write a fraction for each whole number. Draw arrows to show where each fraction or mixed number goes. Show the equations for the fractions and mixed numbers.**

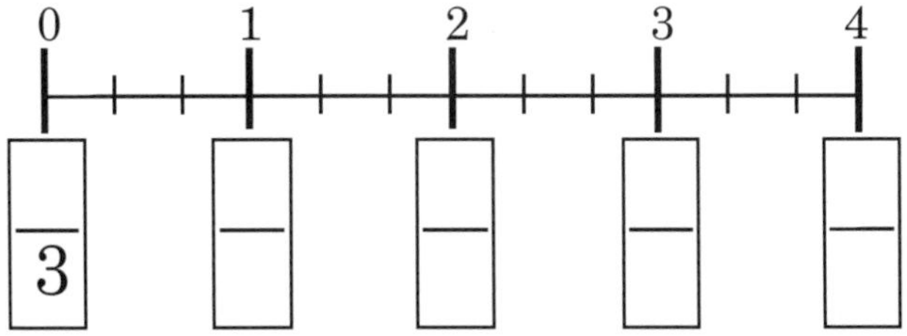


a. $\frac{7}{3}$

b. $1 + \frac{1}{3}$

c. $3 + \frac{2}{3}$

**Part 6** **Make a number family for problems a, b and c. Figure out the answers to all the problems.**

a. During the day, Donna counted 47 birds that were in trees and 21 that were flying. How many birds did she see in all?

b. A truck carried 940 bottles. 310 of them were full. The rest were empty. How many were empty?

c. The tower was 19 feet shorter than the pine tree. The pine tree was 124 feet tall. How tall was the tower?

d. A field is the shape of a rectangle. It is 140 feet long and 23 feet wide. What is the area of the field?

**Part 7** **Copy each problem and work it.**

a. $5\overline{)4700}$ b. $7\overline{)2527}$ c. $4\overline{)832}$ d. $3\overline{)876}$

## Part 8

**Write what $x$ and $y$ equal for each item.**

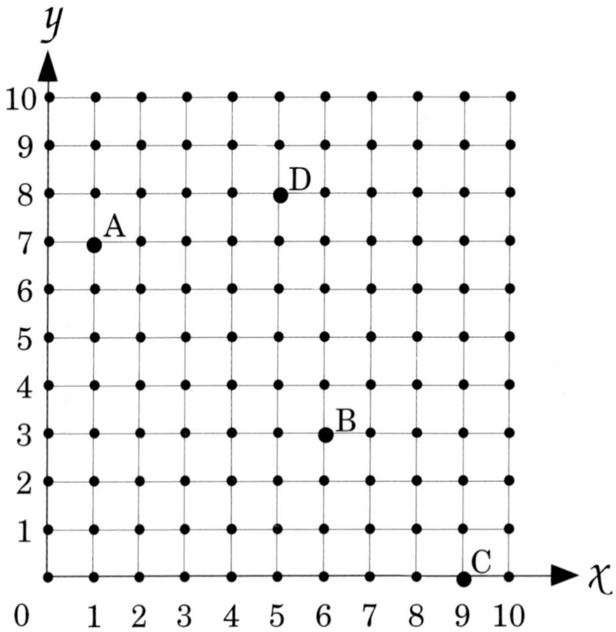


A ($x$= ___, $y$= ___)

B ($x$= ___, $y$= ___)

C ($x$= ___, $y$= ___)

D ($x$= ___, $y$= ___)

## Part J

| | | | |
|---|---|---|---|
| a. | 4 × 9 = 36 | $4\overline{)36}$ = 9 | $\frac{36}{4}$ = 9 |
| b. | 7 × 8 = 56 | $7\overline{)56}$ = 8 | $\frac{56}{7}$ = 8 |
| c. | 6 × 9 = 54 | $6\overline{)54}$ = 9 | $\frac{54}{6}$ = 9 |
| d. | 3 × 5 = 15 | $3\overline{)15}$ = 5 | $\frac{15}{3}$ = 5 |

# Lesson 79

## Part 1

- Here's a problem:

  **Fran had $36. Then her mother paid her $12 for working around the house. She spent some money and ended up with $14. How much did she spend?**

- Here's the number family with **in, out** and **end up.** The number for **end up** is 14.

  **end up** **out** **in**

  **14** ⟶

- You write both values for **in** and add to get a total. The **total** is the big number.

  **end up** **out** **in**

  **36**
  **+ 12**
  **48**

  **14** ⟶ **48**

- You subtract to figure out how much she spent. That's the number for **out.**

  **$48**
  **– 14**
  **$34**

- Fran spent $34.

a. Jan had $17.00 in the bank. Later she put $12.00 in the bank. The next day Jan went to the bank and took out $11.50. How much money did Jan end up with in the bank?

b. A tank had some water in it. 250 gallons were taken from the tank. Then another 720 gallons were taken from the tank. The tank still had 1150 gallons in it. How many gallons were in the tank at the beginning?

c. A farmer had 534 bales of hay. She fed 247 bales to her cattle. She sold 85 bales to a neighbor. She threw 4 bales away because they were rotten. How many bales did she end up with?

## Part 2

- You've worked area problems that tell the lengths of the sides. You multiply the length times the width:

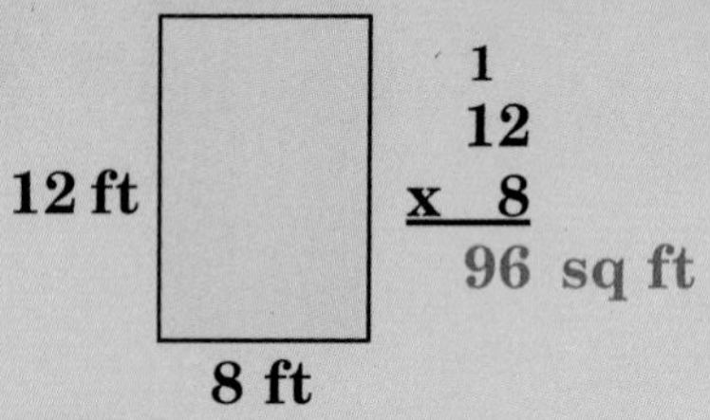


**What is the area of the rectangle?**

- For some problems, you have to figure out the length of a side. Those problems tell the area and tell the length of one side.
- Here's a problem:

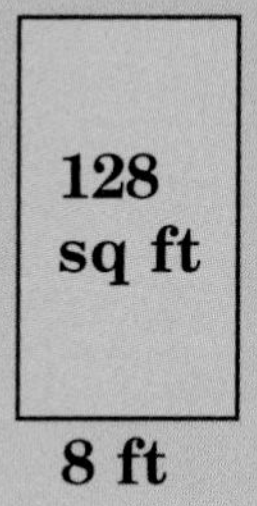


**What is the height of the rectangle?**

- Here's the number problem: **8 x __ = 128**
- Here's the division problem:

**1 6 ft**
**8 ⟌ 1 2 48**
**8**

- The rectangle is 16 feet high.

a.

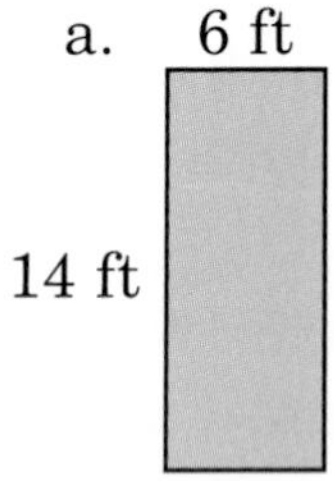


What is the area of the rectangle?

b.

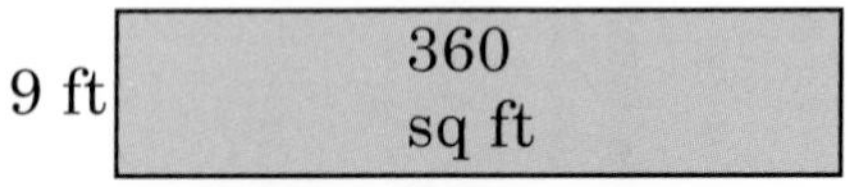


What is the width of the rectangle?

c.

What is the height of the rectangle?

d.

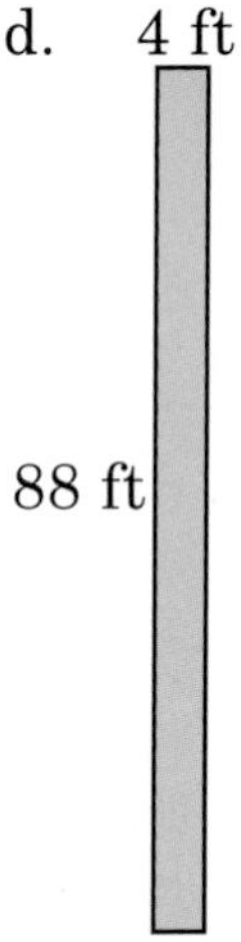


What is the area of the rectangle?

## Part 3

**Sample Problem:** $\frac{2}{7}$ of a rectangle are shaded. The rest is not shaded. What's the fraction for the part that is not shaded?

a. $\frac{2}{3}$ of a circle are shaded. The rest is not shaded. What's the fraction for the part that is not shaded?

b. $\frac{11}{15}$ of a triangle were not shaded. What fraction of the triangle was shaded?

c. $\frac{3}{10}$ of a board were not shaded. What fraction of the board was shaded?

d. $\frac{4}{9}$ of a ruler were shaded. What fraction of the ruler was not shaded?

# Independent Work

## Part 4

**For each row, write the multiplication equation, the division problem and its answer and the fraction equation.**

| | $\square \times \_\_ = \square$ | $\overline{)\phantom{00}}$ | $\frac{\square}{\square} = \square$ |
|---|---|---|---|
| a. | $4 \times \_\_ = 36$ | | |
| b. | | $2\overline{)78}$ | |
| c. | | | $\frac{21}{7} = \square$ |
| d. | | $3\overline{)162}$ | |

## Part 5 Write the names and the equation for the equivalent fractions.

a. There were 12 spokes on each wheel. There were 29 wheels. How many spokes were there?

b. In an orchard, there were 9 apple trees for every 4 fig trees. There were 180 apple trees. How many fig trees were there?

c. For every 3 cones, there were 60 seeds. There were 42 cones. How many seeds were there?

## Part 6 Copy each item and write the mixed number each fraction equals.

a. $\frac{17}{2}$ b. $\frac{46}{3}$ c. $\frac{72}{5}$

## Part 7 Make the number family for each problem. Write the answer to each question.

a. There were 835 trees in the grove. 101 of them were dead. How many were not dead?

b. A chicken farm had 560 red hens and 130 white hens. How many hens were on the farm?

c. Fran was 31 years younger than her mother. Fran was 12 years old. How old was her mother?

d. Mr. Smith was 50 years older than Jill. Mr. Smith was 74 years old. How old was Jill?

That man is 50 years older than the boy. That boy is 10 years old.

# Test 8

## Part 5

a. The numbers for the fraction are 7 and 15. 7 is the denominator.

b. The numbers for the fraction are 12 and 9. The fraction is less than 1.

c. The numbers for the fraction are 183 and 78. The denominator of the fraction is 78.

d. The numbers are 45 and 16. The denominator is 45.

## Part 6

a. 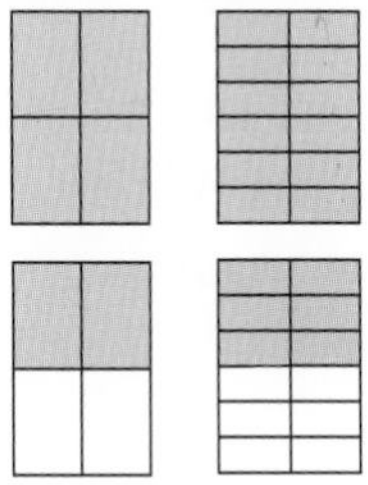

b. 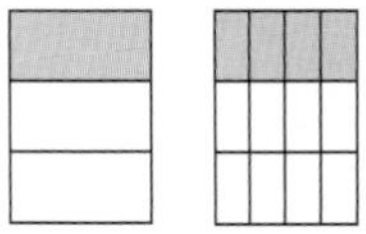

c. 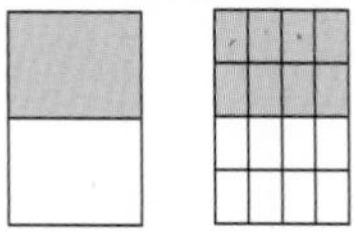

## Part 7

a. Joe had $18.20. Joe's brother gave him $5.00. Joe found $.25 on the ground. He spent some money. Joe ended up with $4.00. How much money did Joe spend?

b. Jan had some money. She spent $7.41. She ended up with $12.50. How much did she have to start with?

c. Jill put some apples into a basket. She gave away 37 of those apples. She ate 4 apples. 17 apples were rotten so she threw them away. She had 14 apples left in the basket. How many apples were in the basket to start out with?

## Part 8

a. $\frac{7}{4} + 3 =$

b. $5 \times \frac{2}{3} =$

c. $\frac{11}{2} \times 6 =$

d. $4 - \frac{4}{1} =$

## Part 9

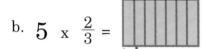

a. In a pond there are 2 bass for every 5 trout. There are 150 trout. How many bass are in the pond?

b. A car burned 2 pints of gas for every 9 miles it traveled. If the car burned 180 pints of gas, how many miles did the car travel?

c. There are 24 bottles in each case. A store has 72 cases. How many bottles are there?

# Lesson 81

## Part 1

a.

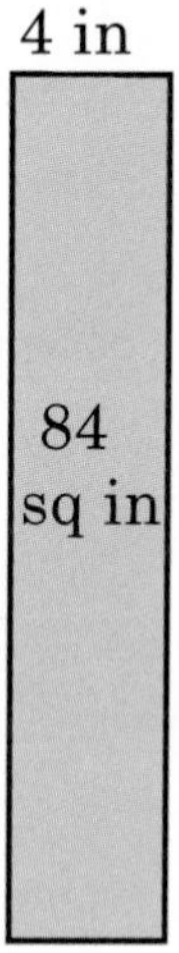


What is the height of the rectangle?

b.

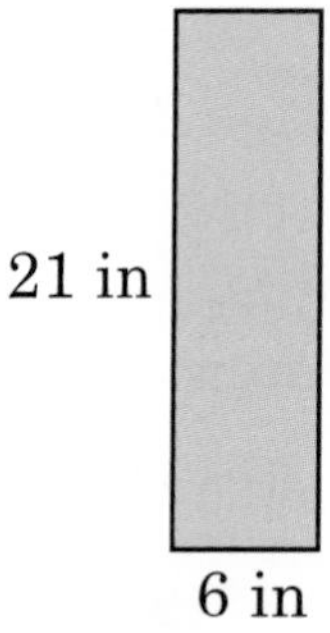


What is the area of the rectangle?

c.

What is the height of the rectangle?

d.

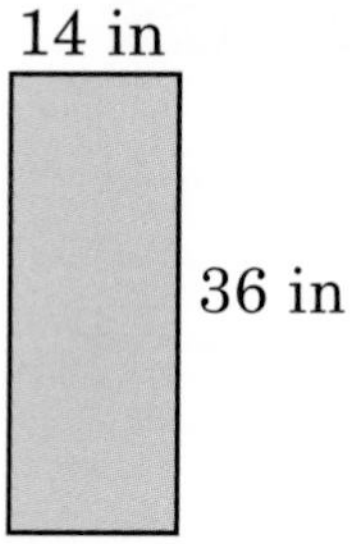


What is the area of the rectangle?

## Part 2

***Sample Problem:*** $\frac{3}{5}$ of the children are girls. What fraction of the children are boys?

a. $\frac{2}{9}$ of the trees in a forest are dead. The rest are alive. What fraction of the trees in the forest are alive?

b. $\frac{6}{11}$ of the days are rainy. The rest of the days are dry. What fraction of the days are dry?

c. In Hill Town, $\frac{5}{8}$ of the days are freezing. What fraction of the days are not freezing?

**Part 3**

a. The bike shop started with 47 bikes. On Monday, the shop sold 14 bikes. On Tuesday, the shop sold 11 bikes. How many bikes were still in the shop?

b. The bike shop started out with lots of bikes. On Monday, the shop sold 56 bikes. On Tuesday, the shop sold another 27 bikes. The shop ended up with 72 bikes. How many bikes did the shop start out with?

c. 678 ants moved into an anthill. Then another 112 ants went into the anthill. Later, some ants left the anthill to go hunting. The anthill ended up with 458 ants. How many ants went hunting?

**Part 4**

***Sample Sentence:*** The ratio of dogs to cats is 7 to 9.

a. The ratio of perch to bass is 3 to 7.

b. The ratio of trees to birds is 2 to 5.

c. The ratio of students to tables is 10 to 3.

## Independent Work

**Part 5** **Copy each mixed number and write the complete equation.**

a. $14 + \frac{1}{4}$ b. $203 + \frac{4}{5}$ c. $90 + \frac{2}{3}$

**Part 6** **Copy each fraction. Complete the equation to show the mixed number each fraction equals.**

a. $\frac{37}{4} =$ b. $\frac{18}{5} =$ c. $\frac{75}{9} =$ d. $\frac{48}{5} =$

## Part 7

**Write the names and a ratio equation for each problem and figure out the answer.**

a. If 5 containers hold 6 pounds of sand, how many pounds of sand will 30 containers hold?

b. In a recipe, there are 6 ounces of water for every 8 ounces of flour. If there are 480 ounces of water, how many ounces of flour are needed?

c. In the kennel, the ratio of cats to dogs was 3 to 5. There were 750 dogs. How many cats were there?

## Part 8

**Copy each problem and work it.**

a. $\frac{2}{3} \times \frac{1}{5} =$

b. $\frac{7}{8} + \frac{1}{8} =$

c. $9 - \frac{2}{8} =$

d. $\frac{2}{5} \times 5 =$

e. $\frac{2}{5} + 5 =$

## Part 9

**Copy the table. Use the information that the facts give to write four numbers in the table. Then fill in the table.**

This table is supposed to show the number of men and women in two different places—Office Plaza and Business Village.

| | Men | Women | Total |
|---|---|---|---|
| Office Plaza | | | |
| Business Village | | | |
| Total for both places | | | |

***Facts***

1. The total number of women at both places is 698.
2. There are 472 men at Office Plaza.
3. The total number of men at both places is 705.
4. At Office Plaza, there are 31 fewer men than women.

## Part 10

**Copy each problem and work it. Check your work with a calculator.**

a. $\begin{array}{r} 127 \\ \underline{\times\ \ 19} \end{array}$

b. $\begin{array}{r} 587 \\ \underline{+\ 291} \end{array}$

c. $\begin{array}{r} 486 \\ \underline{\times\ \ 20} \end{array}$

d. $\begin{array}{r} 127 \\ 86 \\ \underline{+\ 350} \end{array}$

e. $\begin{array}{r} 76 \\ \underline{\times\ 94} \end{array}$

f. $\begin{array}{r} 561 \\ \underline{-\ 462} \end{array}$

g. $\begin{array}{r} 790 \\ \underline{-\ 381} \end{array}$

h. $4\overline{)3408}$

i. $9\overline{)9486}$

j. $3\overline{)9105}$

## Part J

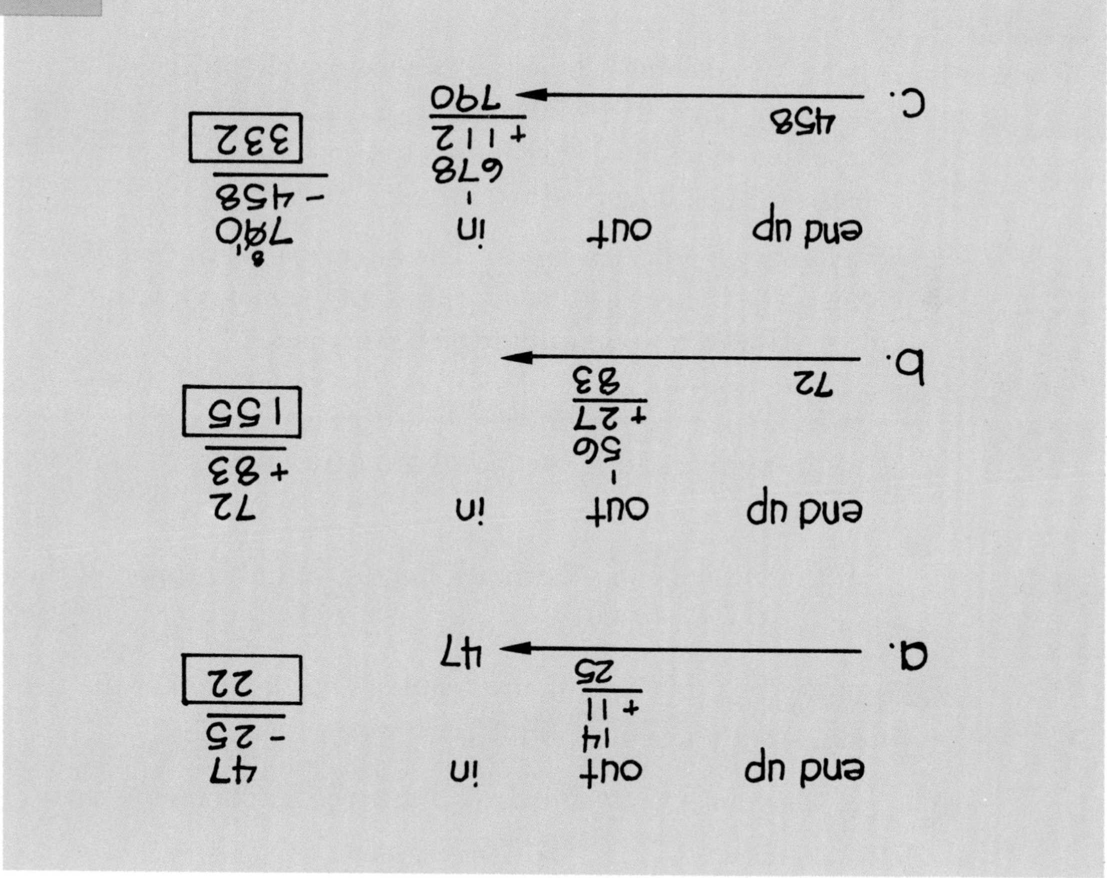

# Lesson 82

## Part 1

a. $\frac{2}{3}$ $\frac{8}{12}$

b. $\frac{2}{5}$ $\frac{10}{30}$

c. $\frac{7}{3}$ $\frac{28}{15}$

d. $\frac{9}{5}$ $\frac{36}{20}$

e. $\frac{1}{10}$ $\frac{20}{200}$

## Part 2

a. A bakery had 159 cookies. The bakery sold 138 cookies. 14 more cookies were eaten. How many cookies did the bakery end up with?

b. A bakery had some cookies. The bakery sold 457 cookies. The bakery gave away 50 cookies. People in the bakery ate 36 cookies. The bakery still had 104 cookies. How many cookies did the bakery start out with?

c. The bakery made 137 cakes. The bakery already had 45 cakes. The bakery sold some cakes and ended up with 64 cakes. How many cakes did the bakery sell?

## Part 3

a. $\frac{31}{40}$ of the vehicles in Motor Town are cars. What fraction of the vehicles are not cars?

b. $\frac{3}{10}$ of the animals on a farm are horses. What fraction of the animals are not horses?

c. $\frac{28}{54}$ of the fish in a stream are trout. What fraction of the fish in the stream are not trout?

d. $\frac{1}{19}$ of the cups are not filled. What is the fraction of the cups that are filled?

## Independent Work

### Part 4

**Copy each mixed number and write the complete equation.**

a. $3 + \frac{2}{5}$    b. $7 + \frac{1}{9}$    c. $65 + \frac{3}{4}$    d. $10 + \frac{5}{18}$

### Part 5

**Write the $x$ and $y$ values for each item.**

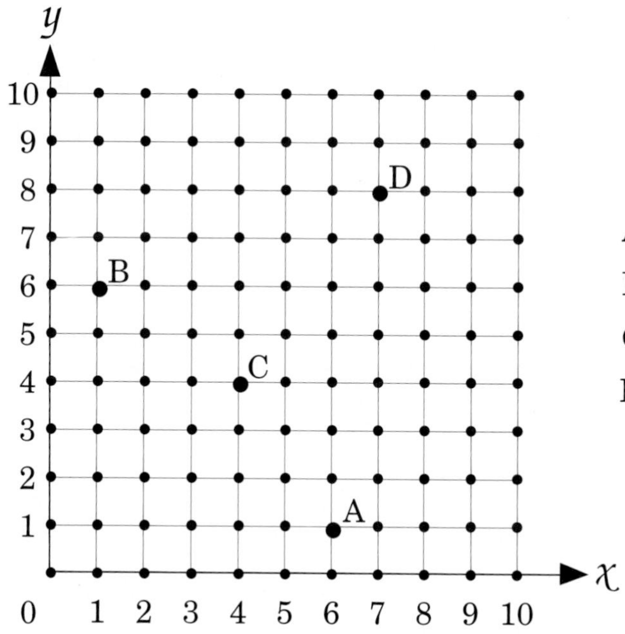


A ($x$= ___, $y$ = ___)

B ($x$= ___, $y$ = ___)

C ($x$= ___, $y$ = ___)

D ($x$= ___, $y$ = ___)

### Part 6

**Write the answer to each problem.**

a. $8\overline{)64}$    b. $7\overline{)63}$    c. $9\overline{)72}$    d. $7\overline{)56}$    e. $9\overline{)54}$

f. $2\overline{)18}$    g. $4\overline{)36}$    h. $6\overline{)36}$    i. $7\overline{)49}$    j. $9\overline{)81}$

## Part 7 **Figure out the area of each rectangle.**

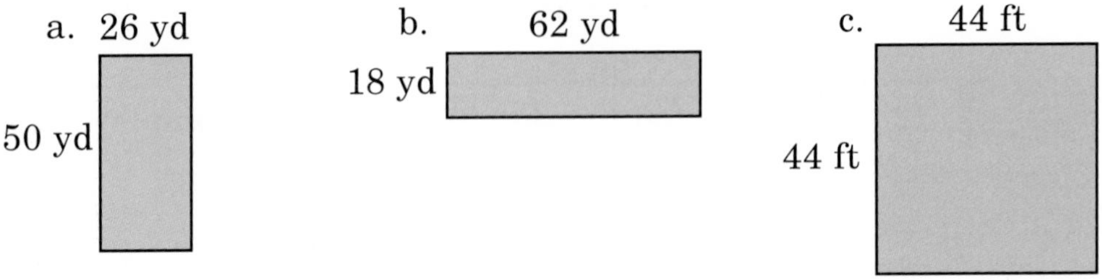


## Part 8 **Make a number family for each problem. Answer the question.**

a. The hog is 121 pounds lighter than the mule. The mule weighs 809 pounds. How much does the hog weigh?

b. There were 12 more birds in the air than in the trees. 41 birds were in the trees. How many were in the air?

c. The rectangle was 314 feet longer than it was wide. The rectangle was 506 feet long. How wide was it?

d. There were 612 towels in a locker room. 47 towels were clean. The rest were dirty. How many towels were dirty?

## Part 9 **Write the names and a ratio equation for each problem and box the answer.**

a. In a bakery, the ratio of cookies to muffins is 3 to 4. There are 120 cookies in the bakery. How many muffins are there?

b. The ratio of windows to doors in a building is 9 to 2. There are 234 doors in the building. How many windows are there?

c. There are 8 nuts in each pack. There are 648 packs. How many nuts are there?

# Lesson 83

## Part 1

a. $\frac{4}{1} \times \blacksquare = 4$

more than 1
= 1
less than 1

b. $7 \times \blacksquare = \frac{7}{1}$

more than 1
= 1
less than 1

c. $7 \times \blacksquare = \frac{1}{5}$

more than 1
= 1
less than 1

d. $\frac{7}{3} \times \blacksquare = \frac{7}{7}$

more than 1
= 1
less than 1

e. $\frac{27}{9} \times \blacksquare = 3$

more than 1
= 1
less than 1

f. $\frac{2}{9} \times \blacksquare = \frac{5}{9}$

more than 1
= 1
less than 1

## Part 2

a. 9 cm — 279 sq cm

b. 4 cm — 84 sq cm

c. 7 mi — 49 mi

d. 7 ft — 168 sq ft

**Part 3**

a. Joe had some money. He spent $4.57 on a book. He loaned his sister $5.00, and he bought a pack of gum for $.36. He ended up with $1.04. How much money did Joe start with?

b. Jill made 137 baskets. She already had 45 baskets. She sold some of the baskets to a market. Jill ended up with 64 baskets. How many baskets did Jill sell?

c. There were 568 carrots growing in a garden. Rabbits ate 43 of the carrots. The gardener pulled up 182 carrots. How many carrots were left in the garden?

**Part 4**

a. A board is 5 feet long. How many inches long is it?

b. How many inches are in 4 yards?

c. A string is 60 feet long. How many yards long is that string?

**Part 5**

a. $\frac{15}{47}$ of the trees in a park were large. What fraction of the trees in the park were not large?

b. $\frac{28}{120}$ of the people in an office were women. What fraction of the people in the office were not women?

c. A cow ate $\frac{72}{90}$ of the hay. Other animals ate the rest of the hay. What fraction of the hay did the other animals eat?

d. Jim scored $\frac{28}{55}$ of the points. What fraction of the points were scored by other players?

## Independent Work

**Part 6** **Write the names and the ratio equation for each problem. Figure out the answer.**

a. There are 4 spots on each bug. There are 360 spots. How many bugs are there?

b. The ratio of flowers to plants is 7 to 2. If there are 614 plants, how many flowers are there?

## Part 7

**For each row, write the multiplication equation, the division problem and its answer and the fraction equation.**

| | $\square \times \_\_ = \square$ | $\overline{)\ \ }$ | $\frac{\square}{\square} = \square$ |
|---|---|---|---|
| a. | | | $\frac{325}{5}$ |
| b. | 7 x __ = 490 | | |
| c. | | $4\overline{)828}$ | |
| d. | | $3\overline{)654}$ | |

## Part 8

**Copy the function table. Then write the function and the answer for each row.**

| | Function | Answer |
|---|---|---|
| $x$ | $x + 8$ | $\square$ |
| 4 | | |
| 0 | | |
| 8 | | |
| 63 | | |
| 47 | | |

## Part 9

**Copy each pair of fractions and write = or ≠.**

a. $\frac{3}{7}$ $\frac{12}{35}$

b. $\frac{9}{2}$ $\frac{63}{14}$

c. $\frac{7}{4}$ $\frac{21}{8}$

## Part 10

**Copy each problem and work it.**

a. $1 - \frac{2}{3} =$

b. $\frac{12}{5} - 1 =$

c. $\frac{9}{10} + \frac{4}{10} =$

d. $\frac{9}{10} \times \frac{4}{10} =$

e. $6 \times \frac{5}{6} =$

f. $\frac{1}{3} \times \frac{0}{5} =$

g. $\frac{4}{9} - \frac{0}{9} =$

## Part 11

**Make a number family for each problem. Answer the question.**

a. Greg worked in March and April. Greg earned \$345 in March. He earned some more money in April. The total he earned for both months was \$579. How much did he earn in April?

b. The bridge weighed 890 pounds less than the truck. The truck weighed 6998 pounds. How much did the bridge weigh?

c. In the morning, they filled 450 more boxes of grapes than they filled in the afternoon. They filled 459 boxes of grapes in the afternoon. How many boxes did they fill in the morning?

d. Rita had some stamps. She got rid of 53 stamps. She ended up with 349 stamps. How many did she start with?

## Part J

a. large / not large / all: 15, 32, 47 → 47 trees

$47 - 15 = 32$

b. women / not women / all: 28, 92, 120 → 120 people

$120 - 28 = 92$

c. cow / other / all: 72, 18, 90 → 90 animals

$90 - 72 = 18$

d. Jim's / others / all: 28, 27, 55 → 55 points

$55 - 28 = 27$

## Part K

a. $304 - 25 = 279$

b. $7006 - 309 = 6697$

c. $2008 - 517 = 1491$

# Lesson 84

## Part 1

a. How many yards are in 165 feet?

b. How many inches are in 130 feet?

c. John walked 9 miles. How many yards did John walk?

## Part 2

a. $\frac{12}{2} \times \blacksquare = 12$

more than 1
= 1
less than 1

b. $\frac{12}{2} \times \blacksquare = 6$

more than 1
= 1
less than 1

c. $\frac{12}{2} \times \blacksquare = 5$

more than 1
= 1
less than 1

d. $\frac{3}{7} \times \blacksquare = \frac{6}{7}$

more than 1
= 1
less than 1

e. $\frac{7}{7} \times \blacksquare = 2$

more than 1
= 1
less than 1

## Part 3

a. There were no people on an elevator. Then 2 people got on the elevator. On the next floor, 5 more people got on the elevator. On the next floor, 4 people got off the elevator. On the next floor, 7 people got on the elevator, and 9 people got off the elevator. How many people were still on the elevator?

b. Jim had 245 baseball cards. He bought 25 more baseball cards. His sister gave him 40 baseball cards. He gave 100 baseball cards to his friend. Then he lost 4 cards. How many baseball cards did Jim end up with?

c. A bike shop had 90 bikes in the storeroom. The shop had 25 bikes on the showroom floor. 18 more bikes were delivered to the shop. The bike shop sold some bikes. The shop had 86 bikes left. How many bikes did the shop sell?

## Part 4

- You can't borrow from a column that has zero. So you go to the first digit that is not zero.

- For this problem, you have to borrow to work in the ones column. You borrow from 60.

$$\begin{array}{r} 600 \\ -131 \\ \hline \end{array}$$

- So you rewrite 60 as 59 plus the 1 you borrow.

$$\begin{array}{r} 59\phantom{0} \\ \cancel{60}{}^{1}0 \\ -131 \\ \hline 469 \end{array}$$

- For this problem, you borrow from 600.

$$\begin{array}{r} 6000 \\ -\ \ 131 \\ \hline \end{array}$$

- So you rewrite 600 as 599 plus the 1 you borrow.

$$\begin{array}{r} 599\phantom{0} \\ \cancel{600}{}^{1}0 \\ -\ \ 131 \\ \hline 5869 \end{array}$$

- For this problem, you have to borrow to work the problem in the tens column.

$$\begin{array}{r} 6003 \\ -\ \ 131 \\ \hline 2 \end{array}$$

- So you borrow from 60. You rewrite 60 as 59 plus the 1 you borrow.

$$\begin{array}{r} 59\phantom{00} \\ \cancel{60}{}^{1}03 \\ -\ \ 131 \\ \hline 5872 \end{array}$$

a. $\begin{array}{r} 704 \\ -109 \\ \hline \end{array}$

b. $\begin{array}{r} 306 \\ -299 \\ \hline \end{array}$

c. $\begin{array}{r} 6003 \\ -\ \ 112 \\ \hline \end{array}$

d. $\begin{array}{r} 7026 \\ -\ \ \ 92 \\ \hline \end{array}$

## Part 5

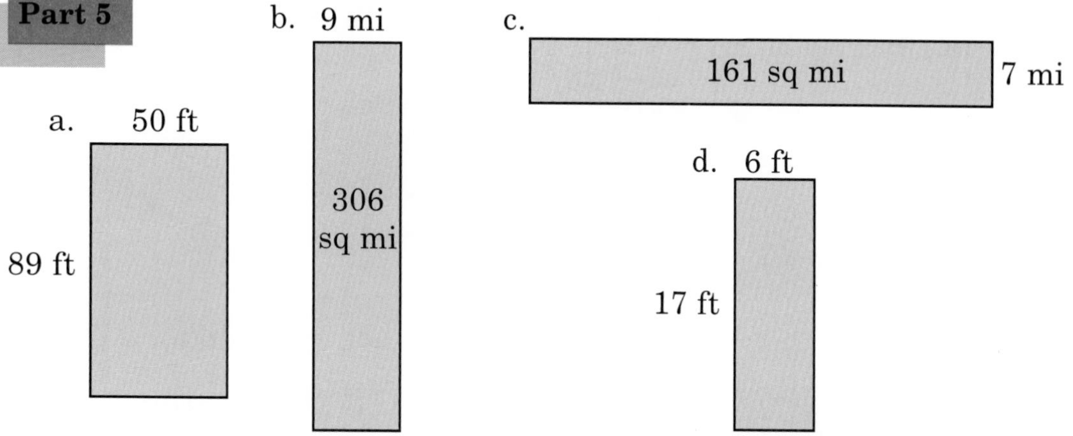


## Part 6

> ***Sample Problem:*** 7 berries are not ripe. 2 berries are ripe. What fraction of the berries are ripe? What fraction of the berries are not ripe?

a. 9 cars are painted. 5 cars are not painted. What fraction of the cars are painted? What fraction of the cars are not painted?

b. There are 73 girls. There are 19 boys. What fraction of the students are girls? What fraction of the students are boys?

c. There are 54 students. 23 are girls. What fraction of the students are boys? What fraction of the students are girls?

## Independent Work

## Part 7

**Copy the function table. Write the function and the answer for each row.**

| | Function | Answer |
|---|---|---|
| $x$ | $x - 3$ | ☐ |
| 3 | | |
| 37 | | |
| 16 | | |
| 9 | | |

## Part 8

**For each problem, make a number family. Then work the number problem.**

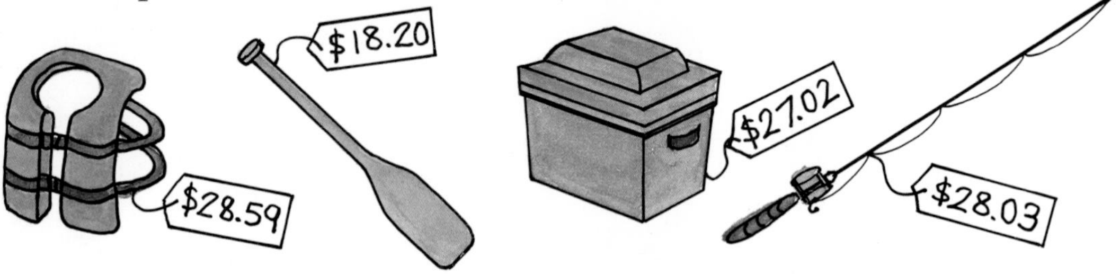


a. The paddle costs $8.82 less than one of the items. Which item is that?

b. Jane bought the life jacket. She received $1.41 in change. How much money did she give the clerk?

c. Sam wants to buy the life jacket. He needs $4.60 more than he has. How much money does he have?

d. Jan has $33.66. She wants to buy the cooler. How much money would she have left?

## Part 9

**Copy each fraction. Complete the equation to show the mixed number each fraction equals.**

a. $\frac{77}{9} =$ b. $\frac{34}{7} =$ c. $\frac{30}{4} =$ d. $\frac{43}{5} =$

## Part 10

**Write the names and the ratio equation for each problem and figure out the answer.**

a. There were 6 cans in each box. There were 312 cans. How many boxes were there?

b. The ratio of cows to horses is 3 to 5. If there are 435 cows, how many horses are there?

**Part 11** **Copy each equation and complete it.**

a. 56 = 9 x ■ + R ■
■

b. 56 = 10 x ■ + R ■
■

c. 65 = 9 x ■ + R ■
■

d. 65 = 8 x ■ + R ■
■

**Part 12** **Copy the number line and write the fraction for each whole number.**

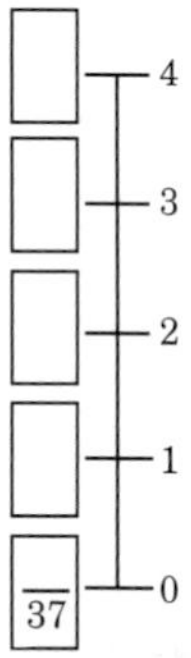


**Part 13** **Copy the table. Figure out all the numbers. Answer the questions.**

This table is supposed to show the number of small rocks and large rocks that are in two parks.

| | Mountain Park | Valley Park | Total for both parks |
|---|---|---|---|
| Small rocks | | | |
| Large rocks | | | |
| Total rocks | | | |

***Facts***

1. In Mountain Park, there are 736 rocks in all.
2. In Mountain Park, there are 372 small rocks.
3. In Valley Park, there are 369 total rocks.
4. There are 204 fewer small rocks in Valley Park than there are in Mountain Park.

***Questions***

a. Are there more large rocks in Mountain Park or in Valley Park?

b. How many large rocks are in both parks?

c. Are there more rocks in Mountain Park or in Valley Park?

d. There are 45 more pine trees than large rocks in Mountain Park. How many pine trees are in Mountain Park?

**Part J**

a. 704 − 109 = 595

b. 306 − 299 = 7

c. 6003 − 112 = 5891

d. 7026 − 92 = 6934

# Lesson 85

## Part 1

- You've learned to find the area of rectangles. The area is the number of squares. For this figure, there are 15 square inches.

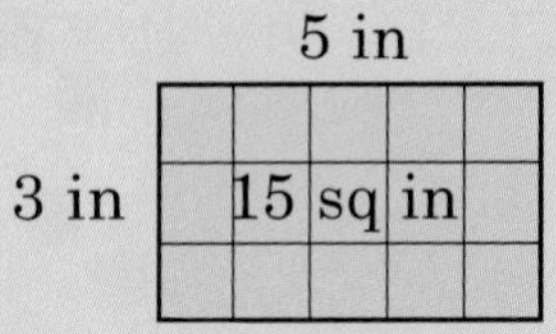


- You can also find the **perimeter** of a rectangle.

**The perimeter is the distance around the rectangle.**

- The dotted arrow shows how far the bug would have to go if it went all the way around the figure. It would have to go along the bottom and then keep going around the rest of the rectangle.

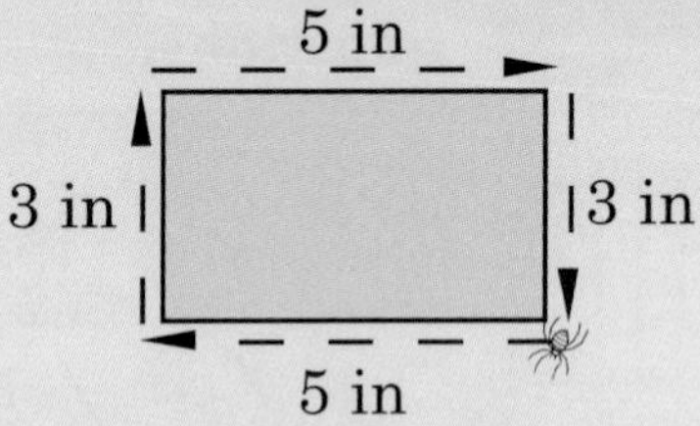


- To find the perimeter, you add up the length of each side. The units are not squares, because you're not going inside the figure to measure the area. You're going along the edges to measure distance.
- Here's the addition problem to find the perimeter of this rectangle:

$$\begin{array}{r} 5 \\ 3 \\ 5 \\ \underline{+\ 3} \\ 16 \text{ in} \end{array}$$

- The units are inches. The perimeter is 16 inches.

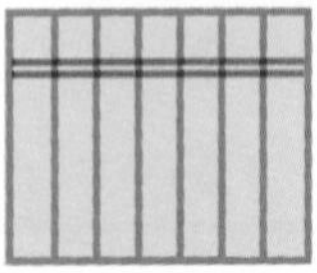

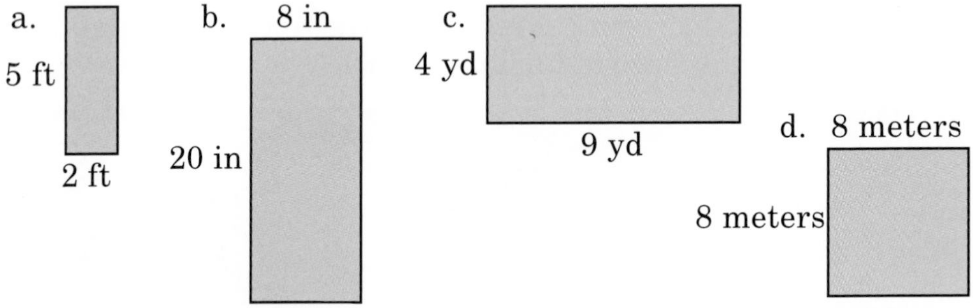

## Part 2

a. James missed 3 items. He got 17 items correct. What fraction of the items did he miss?

b. James worked 60 items. He missed 17 items. What fraction of the items did he get correct?

c. Joe picked 40 apples. 24 of the apples were not ripe. What fraction of the apples were not ripe? What fraction of the apples were ripe?

d. There are 120 doors. Sue painted 83 of the doors. What fraction of the doors did Sue paint? What fraction of the doors were not painted?

## Part 3

a. $\begin{array}{r} 704 \\ -\ 229 \\ \hline \end{array}$

b. $\begin{array}{r} 900 \\ -\ 502 \\ \hline \end{array}$

c. $\begin{array}{r} 7003 \\ -\ \ \ 912 \\ \hline \end{array}$

d. $\begin{array}{r} 6050 \\ -\ \ \ 760 \\ \hline \end{array}$

## Part 4

- You've made points on the coordinate system by using a table that shows $x$ values and $y$ values. When you connect the points, you get a straight line.
- The reason you get a straight line is that all the points follow the same **function rule.**
- Here's a table you've worked with and the line you made:

| | $x$ | $y$ |
|---|---|---|
| A | 5 | 7 |
| B | 1 | 3 |
| C | 7 | 9 |

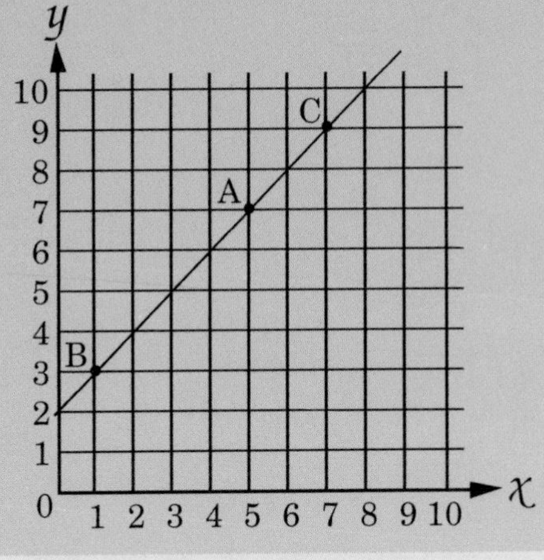


- Here's the same table with the function shown for each item:

| | | Function | Answer |
|---|---|---|---|
| | $x$ | $x + 2$ | $y$ |
| A | 5 | 5 + 2 | 7 |
| B | 1 | 1 + 2 | 3 |
| C | 7 | 7 + 2 | 9 |

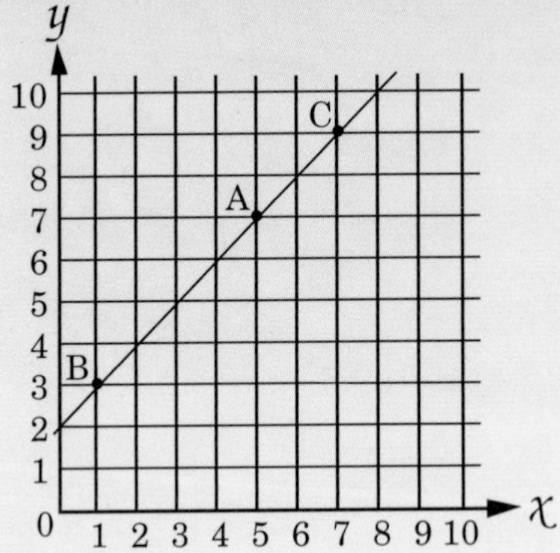


- The function is $x + 2$.

## Part 5

a. On Monday morning, Patti's pottery shop had 147 pots on display. Patti made 46 more pots. She sold some pots. She had 156 pots left at the end of Monday. How many pots did Patti sell?

b. Susan had $147 at first. She got $250 more. Then she got a paycheck for $585. She paid $400 for rent and $249 for a vacation. How much money did Susan have left?

c. Billy had some marbles in a bag. He sold 44 of the marbles. He had 68 marbles left. How many marbles did Billy start with?

d. A pet store had some fish. The store sold 12 neon tetras, 38 goldfish, 64 guppies and 8 swordtails. The store ended up with 634 fish. How many fish did the store start with?

## Part 6

a. Ginger held her breath for 2 minutes. How many seconds was that?

b. How many weeks is 84 days?

c. Tom is 8 years old. How many months old is Tom?

# Independent Work

## Part 7

**For each item, write the multiplication or division problem and figure out the answer.**

a.

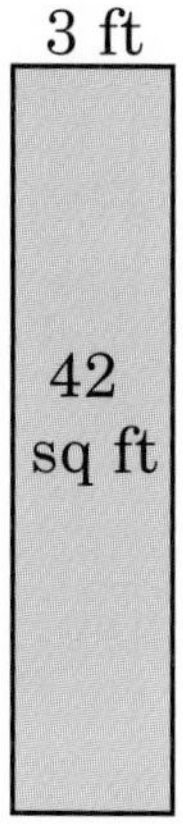


What is the height of the rectangle?

b.

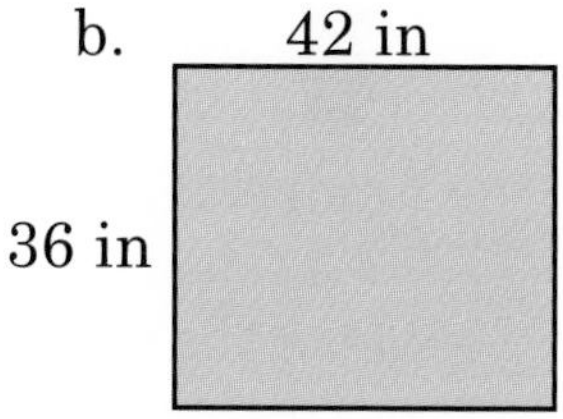

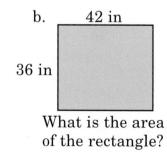

What is the area of the rectangle?

## Part 8

**Make the number family for each problem. Box the answer.**

a. $\frac{3}{10}$ of the melons were soft. The rest were hard. What fraction of the melons were hard?

b. $\frac{5}{8}$ of the dogs were hounds. What fraction of the dogs were not hounds?

## Part 9

**Make a number family for each problem. Answer the question.**

a. There were 450 cars on the lot. 39 of those cars were red. How many cars were not red?

b. There were 121 large birds and 236 small birds in the field. What was the total number of birds in the field?

## Part 10

**Copy each problem and figure out the answer.**

a. $6\overline{)2820}$ b. $4\overline{)8140}$ c. $4\overline{)768}$ d. $4\overline{)352}$

## Part 11

**Write more than 1, less than 1 or = 1 to tell about the missing value.**

a. $\frac{3}{3} \times \blacksquare = \frac{1}{9}$

more than 1
= 1
less than 1

b. $3 \times \blacksquare = \frac{6}{2}$

more than 1
= 1
less than 1

c. $3 \times \blacksquare = \frac{3}{1}$

more than 1
= 1
less than 1

d. $\frac{2}{3} \times \blacksquare = \frac{4}{3}$

more than 1
= 1
less than 1

e. $\frac{1}{7} \times \blacksquare = \frac{4}{7}$

more than 1
= 1
less than 1

f. $1 \times \blacksquare = \frac{4}{5}$

more than 1
= 1
less than 1

## Part 12 Write the ratio equation for each problem. Box the answer.

a. For every 6 windows that were clean, 2 windows were dirty. There were 240 clean windows. How many dirty windows were there?

b. The ratio of tall trees to short trees was 2 to 17. There were 160 tall trees. How many short trees were there?

## Part J

c. apples

| | not ripe | ripe | all |
|---|---|---|---|
| | 24/40 | 16/40 | 40/40 |

d. doors

| | painted | not painted | all |
|---|---|---|---|
| | 83/120 | 37/120 | 120/120 |

## Part K

a. 704 − 229 = 475

b. 900 − 502 = 398

c. 7003 − 912 = 6091

d. 6050 − 760 = 5290

## Part L

a. end up / out / in: 156 → 193

147 + 46 = 193; 193 − 156 = 37

b. end up / out / in: → 982

400 + 249 = 649; 147 + 250 + 585 = 982; $982 − 649 = $333

c. end up / out / in: 68, 44 →

68 + 44 = 112

d. end up / out / in: 634, 122 →

12 + 38 + 64 + 8 = 122; 634 + 122 = 756

# Lesson 86

## Part 1

a. There were 82 ships in the harbor. 31 of the ships were empty. What fraction of the ships were not empty?

b. 71 workers were in the union. 63 workers were not in the union. What fraction of the workers were in the union?

c. 25 pairs of shoes were sneakers. 16 pairs were not sneakers. What fraction of the shoes were not sneakers?

d. 28 pairs of Sue's shoes were not sneakers. Sue had 30 pairs of shoes. What fraction of Sue's shoes were sneakers?

## Part 2

a. A tub contained 64 quarts. How many gallons were in the tub?

b. A dog drank 32 cups of water. How many tablespoons of water did the dog drink?

c. How many quarts are in 54 pints?

## Part 3

| | |
|---|---|
| Mow the lawn | $ 5.00 |
| Clean garage | $ 7.00 |
| Babysit | $ 8.00 |
| Sell your bike | $40.00 |
| Win a prize | $ 9.00 |
| Receive allowance | $12.00 |

***Sample Problem:*** Henry needs money to buy some things. He mows the lawn and sells his bike. With the money he receives, he buys a tape and running shoes. How much money does he have left?

a. Janice got some money from winning a prize. She got some more money for her allowance. She earned money by cleaning the garage and babysitting. She bought a sweatshirt. How much money did Janice end up with?

b. A tub had 302 gallons of water in it. 59 gallons went into the tub when it rained. A man poured another 27 gallons of water into the tub. Some water leaked out of the tub. There were 197 gallons left in the tub. How much water leaked out of the tub?

c. Debbie had some money. She bought a skateboard and a tape, and she went to a movie. She had $16 left. How much money did Debbie start with?

## Part 4

a. $\begin{array}{r} 306 \\ -\underline{129} \end{array}$ b. $\begin{array}{r} 400 \\ -\underline{292} \end{array}$ c. $\begin{array}{r} 9003 \\ -\underline{783} \end{array}$

## Part 5

a.

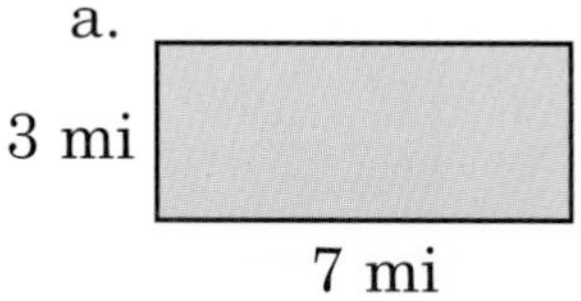


b.

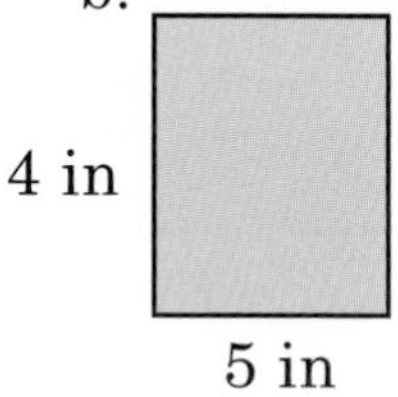


c.

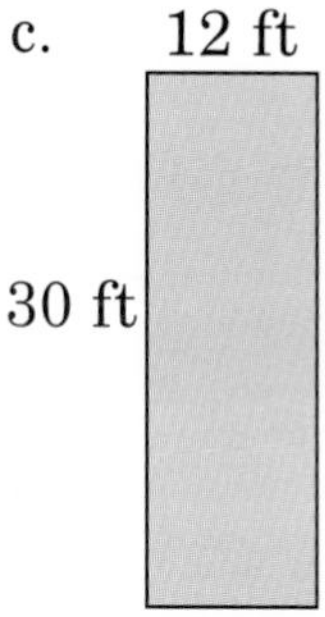


## Independent Work

## Part 6

**Write the multiplication problem for each rectangle and answer the question.**

a.

What's the height of the rectangle?

b.

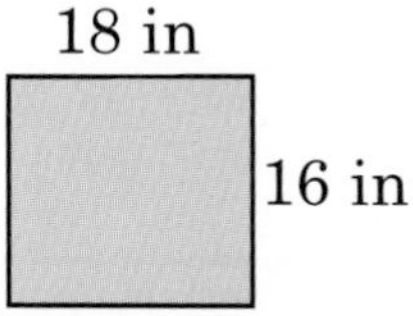


What's the area of the rectangle?

c.

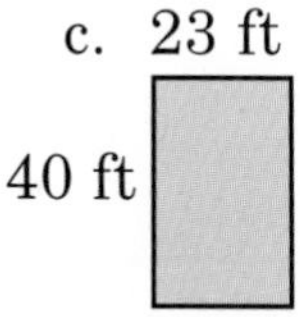


What's the area of the rectangle?

d.

What's the width of the rectangle?

## Part 7

**For each problem, make a number family. Figure out the missing number and box it.**

a. $\frac{15}{28}$ of the lights in a building were on. What fraction of the lights in the building were not on?

b. Sara had money in her pocket and in a jar. She had $\frac{45}{67}$ of her money in the jar. What fraction of Sara's money was in her pocket?

c. In the morning, Tim ran 165 meters. In the afternoon, Tim ran 561 meters. How many more meters did Tim run in the afternoon than in the morning?

d. Carla has 56 more stamps than Fran has. If Fran has 213 stamps, how many stamps does Carla have?

## Part 8

**For items a–f, write the fraction. Copy items g–i and complete the equations.**

a.
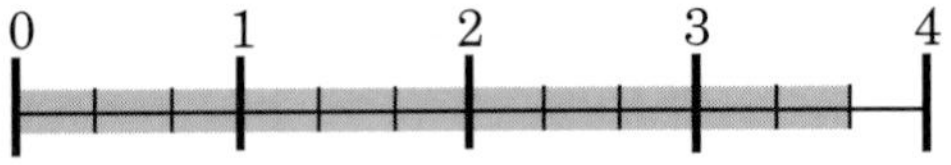


b.
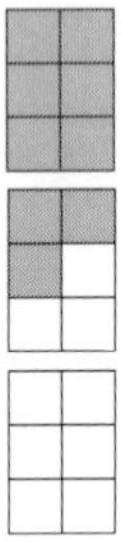

c.
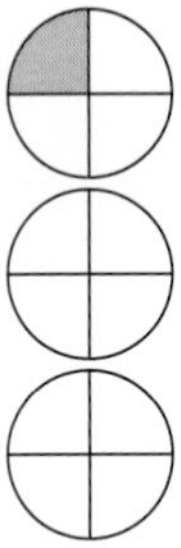

d. The numerator is 75. The denominator is 12.

e. There are 16 parts in each unit. 12 parts are shaded.

f. The numbers for a fraction are 100 and 53. The fraction is less than 1.

g. $\frac{\quad}{25} = 7$

h. $3 = \frac{\quad}{7} = \frac{\quad}{2} = \frac{\quad}{9}$

i. $\frac{14}{3} \times \frac{1}{7} =$

## Part 9

**Copy each problem and work it.**

a. $\frac{12}{2} - 3$

b. $4 + \frac{1}{8}$

c. $\frac{21}{5} + \frac{18}{5}$

**Part J**

b.

| | union | not union | all |
|---|---|---|---|
| workers | 71 | 63 | 134 |
| shoes | 134 | 134 | 134 |

c.

| | sneakers | not sneakers | all |
|---|---|---|---|
| | 25 | 16 | 41 |
| shoes | 41 | 41 | 41 |

d.

| | not sneakers | sneakers | all |
|---|---|---|---|
| | 28 | 2 | 30 |
| shoes | 30 | 30 | 30 |

**Part K**

b. end up out in

197 + 27 → 388

$\overset{2}{3}\overset{1}{8}8 - 197 = $ **191 gallons**

302 − 59

c. end up out in

16 62

45 − 11 + 6

$16 + $62 = **$78**

**Part L**

a. 306 − 129 = 177

b. 400 − 292 = 108

c. 9003 − 783 = 8220

# Lesson 87

## Part 1

- You know that this value tells about dollars and cents: $4.25
- The **4** tells how many **whole** dollars. $**4**.25
- The **25** tells how many **cents.** The cents are the smaller units. $4.**25**
- There are numbers that tell about whole numbers and parts. These numbers work a lot like dollars and cents. They have a decimal point. 4.25
- The number before the decimal point is the whole number. The two digits after the decimal point tell about hundredths.
- Here is 9 and 37 hundredths: That's 9 whole units and 37 tiny parts of another unit. 9.37

| | | |
|---|---|---|
| $7.25 | $13.04 | $7.10 |
| 7.25 | 13.04 | 7.10 |

## Part 2

***Sample Problem:*** Write the number for 1 and 42-hundredths.

a. Write the number for 6 and 30-hundredths.

b. Write the number for 7 and 91-hundredths.

c. Write the number for 18-hundredths.

d. Write the number for 5 and 70-hundredths.

e. Write the number for 1 and 18-hundredths.

## Part 3

a. $\frac{1}{7}$ of the people were wealthy. What fraction of the people were not wealthy?

b. 12 bottles were not broken. 3 bottles were broken. What fraction of the bottles were broken?

c. James missed 4 items. He got the rest of the items correct. There were 25 items in all. What fraction of the items did he miss?

d. $\frac{5}{8}$ of the students in the class were boys. What fraction of the students were girls?

## Part 4

a. $\begin{array}{r} 805 \\ -\ 226 \\ \hline \end{array}$

b. $\begin{array}{r} 3000 \\ -\ 2790 \\ \hline \end{array}$

c. $\begin{array}{r} 307 \\ -\ 199 \\ \hline \end{array}$

d. $\begin{array}{r} 3004 \\ -\ 1068 \\ \hline \end{array}$

## Part 5

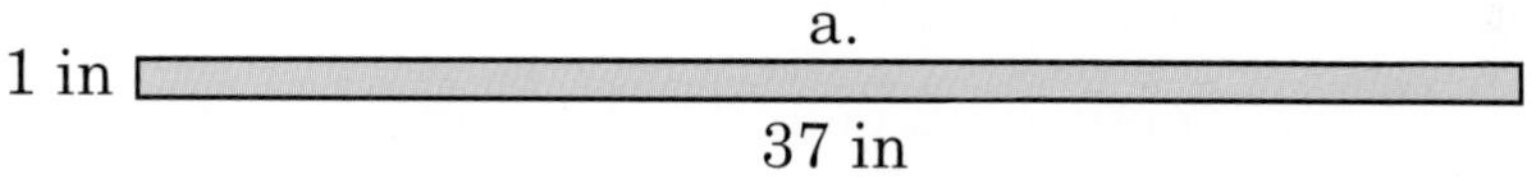


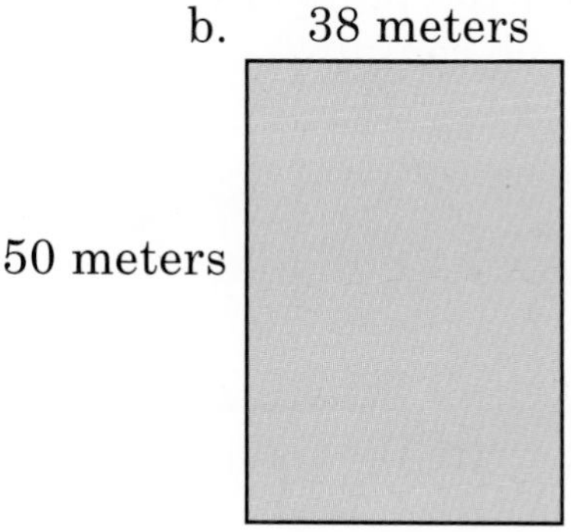

**Part 6**

| | |
|---|---|
| Mowing the lawn | \$7.00 |
| Babysitting | \$3.50 |
| Mopping the floor | \$2.00 |
| Washing the car | \$3.00 |
| Winning a prize | \$5.00 |
| Allowance | \$2.25 |

| ***Menu*** | |
|---|---|
| Salad | \$2.00 |
| Hamburger | \$2.60 |
| Cheeseburger | \$2.90 |
| Juice | \$ .80 |
| Milk | \$1.10 |
| Steak | \$5.30 |
| Fries | \$1.20 |

a. Mary got money from mowing the lawn, from babysitting and from winning a prize. She went to the restaurant and bought dinner. She ordered a hamburger, a salad and a glass of milk. How much money did she end up with after she paid for her meal?

b. Debbie had some money. She bought a steak, some fries and juice. She had \$.64 left. How much did Debbie have to start with?

c. Joe had \$3.10 in his pocket. He earned some more money by mopping the floor and washing the car. Joe bought a hamburger and juice. How much money did Joe have left?

**Part 7**

a. How many weeks are in 4 years?

b. A dog weighed 32 pounds. How many ounces did the dog weigh?

c. A tub contained 168 pints of water. How many quarts of water did the tub contain?

## Independent Work

**Part 8** **For each problem, write the names and the ratio equation. Box the answer.**

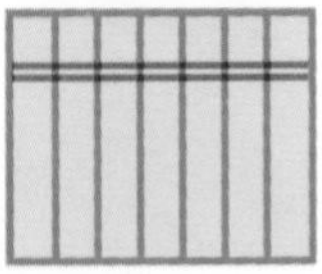

a. The ratio of animals to cages is 9 to 4. There were 108 cages in the zoo. How many animals were in the zoo?

b. For every 5 shelves John builds, he uses 6 bags of nails. If John builds 120 shelves, how many bags of nails would he use?

## Part 9

**Write more than 1, less than 1 or = 1 for each problem.**

a. $\frac{4}{7} \times \blacksquare = \frac{6}{7}$

more than 1
= 1
less than 1

b. $\frac{24}{6} \times \blacksquare = \frac{10}{10}$

more than 1
= 1
less than 1

c. $16 \times \blacksquare = \frac{9}{2}$

more than 1
= 1
less than 1

## Part 10

**Copy the table. Use the information that the facts give to write four numbers in the table. Fill in the table. Then answer the questions.**

This table is supposed to show the number of green ducks and black ducks that were on two ponds.

| | Mill Pond | Flint Pond | Total for both ponds |
|---|---|---|---|
| Green ducks | | | |
| Black ducks | | | |
| Total ducks | | | |

***Facts***

1. The total ducks for both ponds was 480.
2. There were 163 black ducks on Mill Pond.
3. There were 75 green ducks on Mill Pond.
4. There were 12 more green ducks on Flint Pond than on Mill Pond.

***Questions***

a. On Mill Pond, were there more black ducks or green ducks?

b. Which pond had more ducks?

c. How many green ducks were on Flint Pond?

d. On Flint Pond, there were 24 more geese than green ducks. How many geese were on Flint Pond?

## Part 11

**Write each problem as a division problem and figure out the answer.**

a. 7 x ■ = 357  b. 5 x ■ = 305  c. 9 x ■ = 180

d. $\frac{720}{8}$ =  e. $\frac{426}{6}$ =  f. $\frac{355}{5}$ =

## Part J

a. wealthy $\frac{1}{7}$, not wealthy $\frac{6}{7}$, all $\frac{7}{7}$ — people

b. not broken $\frac{12}{15}$, broken $\frac{3}{15}$, all $\frac{15}{15}$ — bottles

c. missed $\frac{4}{25}$, correct $\frac{21}{25}$, all $\frac{25}{25}$ — items

d. boys $\frac{5}{8}$, girls $\frac{3}{8}$, all $\frac{8}{8}$ — students

## Part K

a. 805 − 226 = 579

b. 3000 − 2790 = 210

c. 307 − 199 = 108

d. 3004 − 1068 = 1936

## Part L

a. end up, out, in: 5.30 − 1.20 + .80 = 7.30; $.64 + 7.30 = $7.94

b. end up, out, in: 2.60 + .80 = 3.40; 3.10 + 2.00 + 3.00 = 8.10; $8.10 − 3.40 = $4.70

# Lesson 88

**Part 1**

a. There were some people in a large room. 21 people left the room. Then 11 more people left the room. There were still 41 people in the room. How many people were in the room at first?

b. There were 14 more men than women in a room. There were 71 men in the room. How many women were in the room?

c. During the morning, 75 people were in a large room. During the evening, there were 56 people in the room. How many more people were in the room during the morning than during the evening?

d. A large room had 4 people in it. Then 12 people came into the room. Then 9 people left the room. Then 34 people came into the room. Then 7 people left the room. How many were still in the room?

**Part 2**

a. 12 and 3-hundredths

b. 7 and 9-hundredths

c. 12-hundredths

d. 6-hundredths

e. 300 and 24-hundredths

**Part 3**

***Sample Problem:*** There were 45 trees in a park. 13 trees were dead. How many trees were not dead? What fraction of the trees were dead?

a. There were 56 moths in a yard. 13 were blue. How many moths were not blue? What fraction of the moths were blue?

b. On Monday, $\frac{7}{41}$ of the children were sick. The rest were not sick. How many children were there in all? What fraction of all the children were not sick?

c. Cindie's Kennel was taking care of 65 pets. 14 of the pets did not have collars. How many pets had collars? What's the fraction of pets that did not have collars?

**Part 4**

a. $\frac{42}{7} =$ b. $\frac{30}{6} =$ c. $\frac{54}{6} =$ d. $\frac{48}{6} =$

e. $\frac{18}{6} =$ f. $\frac{18}{9} =$ g. $\frac{56}{7} =$

**Part 5**

- You can find the perimeter for any figure that has straight sides. You just add up the length of each side.

- Here's a figure with five sides:
- You add the length of each side:

**4 + 6 + 3 + 5 + 2 = 20 inches**

The perimeter is 20 inches.

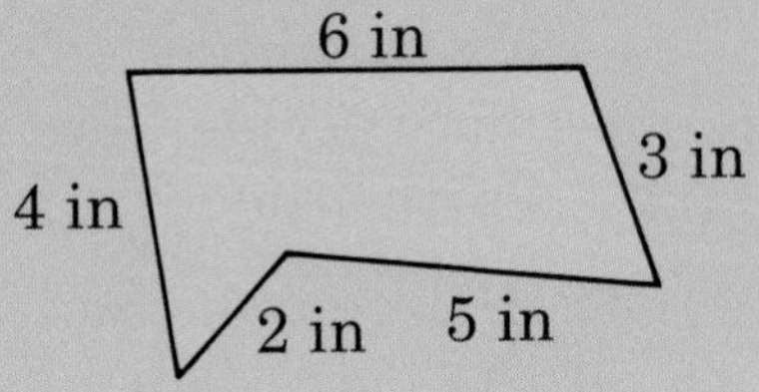


- Here's a triangle:
- You add the length of each side:

**3 + 4 + 5 = 12 inches**

The perimeter is 12 inches.

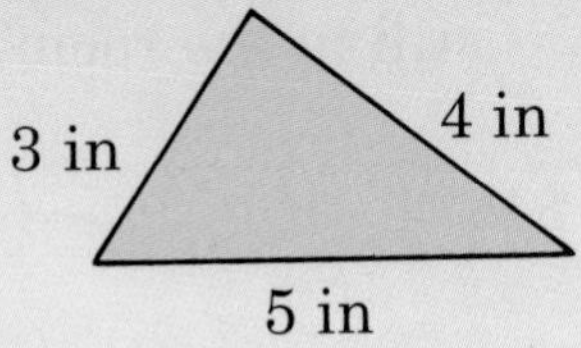


a.

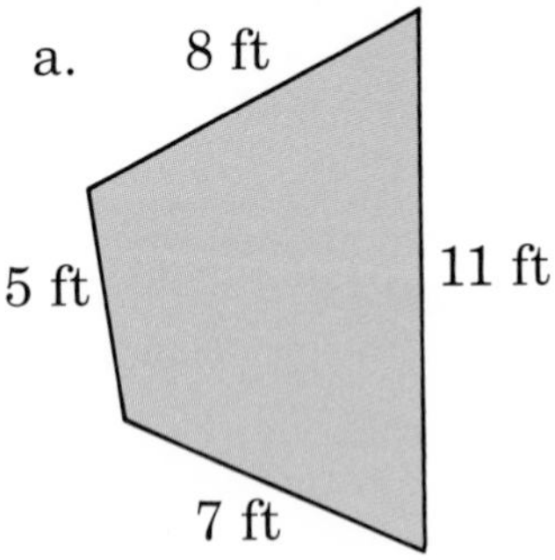


b.

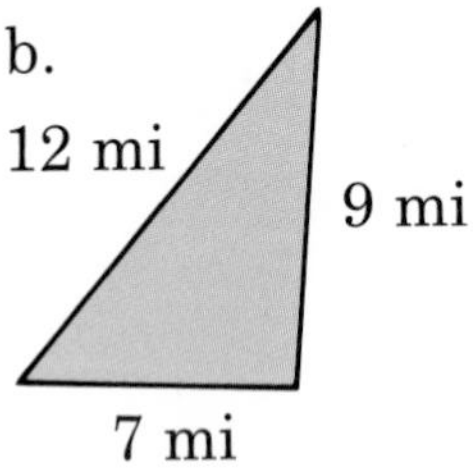


c.

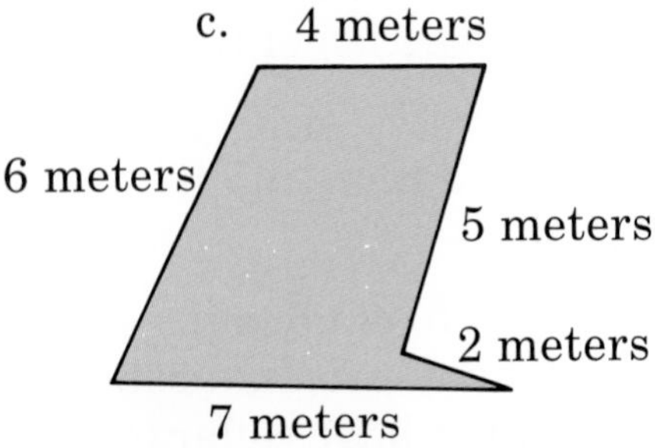


d.

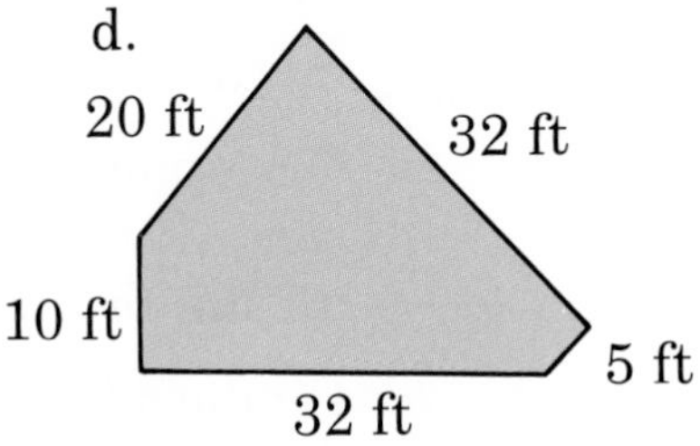

**Part 6**

a. Jim went on vacation for 84 days. How many weeks was Jim on vacation?

b. A baker needed 84 cups of milk to make cakes. How many quarts of milk did the baker need?

c. A boy ran 2 miles. How many feet did the boy run?

## Independent Work

**Part 7** **Make a number family for each problem. Box the answer to each question.**

a. Jim had some money. He bought the gloves and the helmet. He ended up with $48.05. How much money did he start out with?

b. Celia had $64.99. She received $11.00 from her brother. Then she bought one of the items. She ended up with $53.14. How much did Celia spend? Which item did Celia buy?

**Part 8** **Write the column problem for each item and work it.**

a. 4001 – 68 = ■ b. 7082 x 15 = ■ c. 7082 – 95 = ■

d. 1806 + 97 = ■ e. 6002 – 41 = ■

**Part 9** **Copy each fraction and write the mixed number it equals.**

a. $\frac{74}{9}$ b. $\frac{18}{4}$ c. $\frac{8}{7}$

**Part 10** **Copy the fractions for each item. Write = or ≠ between each pair of fractions. (Hint: If you multiply the fraction you start with by 1, the fractions are equal.)**

a. $\frac{7}{9}$ $\frac{63}{72}$ b. $\frac{2}{1}$ $\frac{12}{6}$ c. $\frac{5}{3}$ $\frac{15}{12}$

**Part 11** **Copy each problem and work it.**

a. $\frac{2}{7} \times \frac{4}{4} =$ b. $\frac{2}{7} + \frac{2}{7} =$ c. $\frac{2}{7} \times \frac{7}{2} =$ d. $\frac{3}{5} - \frac{3}{5} =$

**Part 12** **For each problem, write a ratio equation and box the answer.**

a. The ratio of leaves on the ground to all the leaves is 5 to 9. If there are 765 leaves in all, how many leaves are on the ground?

b. In Joe's Restaurant, there are 3 women to every 4 men. How many men are in the restaurant if there are 63 women?

c. In Rosie's Cafe, the ratio of customers to workers is 5 to 2. If there are 70 customers, how many workers are in Rosie's Cafe?

d. In Newton School, there are 4 girls to every 7 students. There are 147 students in Newton School. How many girls go to Newton School?

# Lesson 89

## Part 1

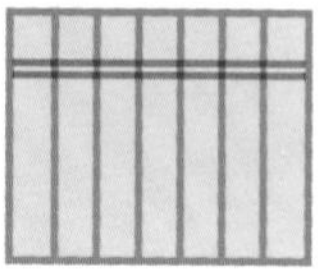

a. There were 9 books on a shelf. 7 books were removed from the shelf. Then 3 books were placed on the shelf. Then 21 more books were placed on the shelf. Then 12 books fell off the shelf. How many books ended up on the shelf?

b. There were red books and blue books on a shelf. There were 9 more red books than blue books on the shelf. If there were 62 blue books on the shelf, how many red books were on the shelf?

c. A shelf was empty. 12 books were placed on the shelf. Then 72 more books were placed on the shelf. Some books were removed from the shelf. There were still 11 books on the shelf. How many books were removed?

d. Some books fell off a shelf. After that happened, there were 31 fewer books on the floor than on the shelf. If there were 15 books on the floor, how many books were still on the shelf?

## Part 2

a. How many centimeters are in 2 meters?

b. How many meters are in 2 kilometers?

c. How many centimeters are in 90 millimeters?

## Part 3

- You can multiply money amounts.
- Here's \$3.56 x 4:

$$\begin{array}{r} \$\ 3.56 \\ \underline{\times\qquad 4} \end{array}$$

- To figure the answer, you multiply. You put a dollar sign in the answer. Then you put a decimal point before the last two digits.

$$\begin{array}{r} {\scriptstyle 2\ 2\ } \\ \$\ 3.56 \\ \underline{\times\qquad 4} \\ \$14.24 \end{array}$$

a. $\begin{array}{r} \$6.11 \\ \underline{\times\quad 6} \end{array}$

b. $\begin{array}{r} \$12.03 \\ \underline{\times\qquad 7} \end{array}$

c. $\begin{array}{r} \$5.67 \\ \underline{\times\ \ 28} \end{array}$

d. $\begin{array}{r} \$2.09 \\ \underline{\times\ \ 19} \end{array}$

## Part 4

a.

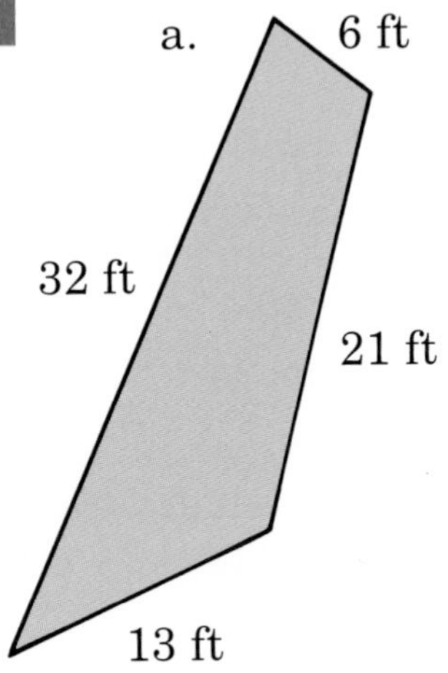


b.

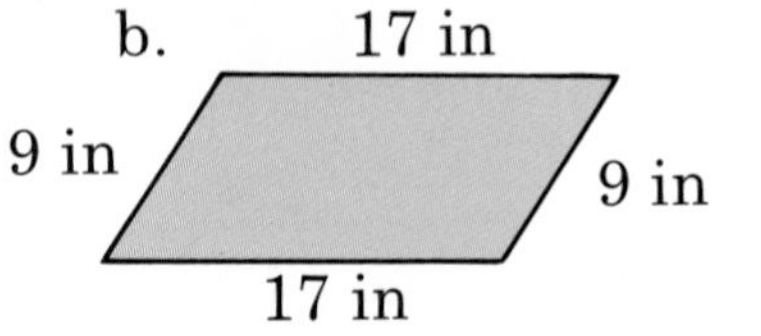


c.

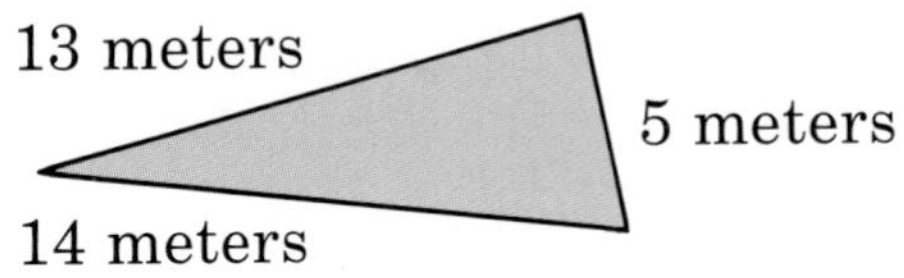


## Part 5

- You can take any decimal number and write it as a fraction.
- Here's 72-hundredths: **.72**
- The name **hundredths** tells you that the denominator of the fraction is 100.
- Here's the fraction for 72-hundredths: $\mathbf{\frac{72}{100}}$
- Here's 6-hundredths: **.06**
- The name hundredths tells you that the denominator of the fraction is 100.
- Here's the fraction for 6-hundredths: $\mathbf{\frac{6}{100}}$
  Notice that the top number is 6, **not zero 6.**
- Here's the decimal number for 3 and 24-hundredths: **3.24**
- Here's the fraction: $\mathbf{\frac{324}{100}}$

a. .06 =    b. 5.40 =    c. .64 =    d. .01 =

## Part 6

a. There were 97 cars on the road. 3 were out of gas. How many cars were not out of gas? What fraction of the cars were out of gas?

b. 14 cans of paint were open. The rest were unopened. There were 62 cans in all. What fraction of the cans were open? What fraction of the cans were unopened?

c. 19 dogs were hungry. 37 dogs were not hungry. How many dogs were there? What fraction of the dogs were hungry?

d. 11 players were injured. There were 56 players. How many were not injured? What fraction of the players were injured?

# Independent Work

## Part 7

**Copy the number line and write the fractions for the whole numbers.**

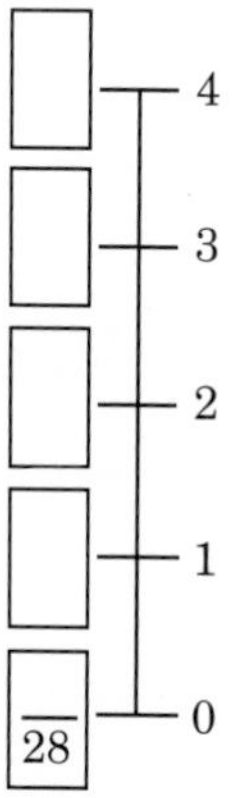


## Part 8

**Answer each question.**

a. $8 + \frac{5}{6}$ = what fraction?

b. $\frac{75}{9}$ = what mixed number?

c. $\frac{13}{5}$ = what mixed number?

d. $362 + \frac{1}{4}$ = what fraction?

## Part 9

**Write more than 1, = 1 or less than 1 to tell about the missing number in each item.**

a. $7 \times \blacksquare = \frac{21}{3}$

more than 1
= 1
less than 1

b. $\frac{16}{7} \times \blacksquare = \frac{8}{7}$

more than 1
= 1
less than 1

c. $\frac{1}{3} \times \blacksquare = \frac{10}{10}$

more than 1
= 1
less than 1

## Part 10

**Make the number family for each item. Box the answers.**

a. There were 75 fewer red birds than blue birds. There were 89 red birds. How many blue birds were there?

b. $\frac{1}{4}$ of the birds were blue birds. What fraction of the birds were not blue birds?

## Part 11

**Copy the problems and figure out the answers.**

a. $7\overline{)476}$  b. $6\overline{)6084}$  c. $7\overline{)7455}$

## Part 12

**Copy the fractions for each item. Figure out the fraction you multiply the first fraction by to get the second fraction. Then write = or ≠ between the fractions for the item.**

a. $\frac{7}{3}$ $\frac{49}{21}$  b. $\frac{2}{9}$ $\frac{30}{135}$  c. $\frac{6}{5}$ $\frac{246}{215}$

## Part J

b. $\frac{\text{kilometers}}{\text{meters}}$ $\frac{1}{1000} \times \frac{2}{2} = \frac{2}{\boxed{2000}}$

c. $\frac{\text{centimeters}}{\text{millimeters}}$ $\frac{1}{10} \times \frac{9}{9} = \frac{\boxed{9}}{90}$

## Part K

b. $12.03 × 7 = $84.21

c. $5.67 × 28 = 45 36 + 113 40 = $158.76

d. $2.09 × 19 = 18 81 + 20 90 = $39.71

## Part L

a.

| out | not out | all |
|---|---|---|
| 3 | 94 | 97 |
| 97 | 97 | 97 |

cars

b.

| open | unopened | all |
|---|---|---|
| 14 | 48 | 62 |
| 62 | 62 | 62 |

cans

c.

| hungry | not hungry | all |
|---|---|---|
| 19 | 37 | 56 |
| 56 | 56 | 56 |

dogs

d.

| injured | not injured | all |
|---|---|---|
| 11 | 45 | 56 |
| 56 | 56 | 56 |

players

# Test 9

**Part 5**

a. The numbers for the fraction are 6 and 8. The numerator is 8.

b. The numbers for the fraction are 17 and 12. The fraction is less than 1.

c. The denominator is 26. The numerator is 50.

d. The numbers for the fraction are 100 and 13. The numerator is 100.

**Part 6**

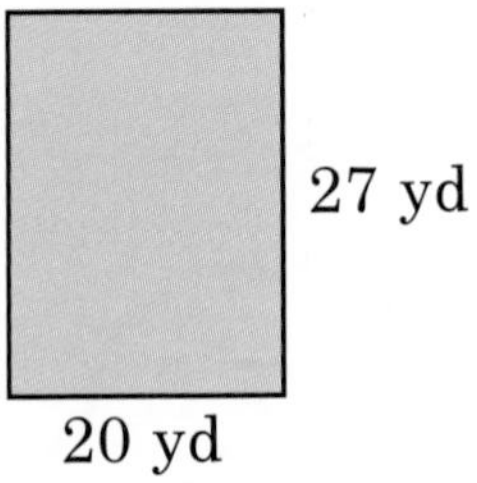


**Part 7**

a. $\frac{14}{30}$ of the leaves were brown. What fraction of the leaves were not brown?

b. Cats and dogs were in a kennel. There were 18 cats and 16 dogs. What is the fraction for all pets that were in the kennel?

**Part 8**

a.

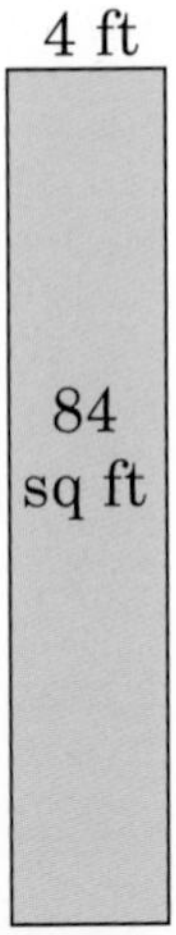


What's the height of the rectangle?

b.

What's the width of the rectangle?

c.

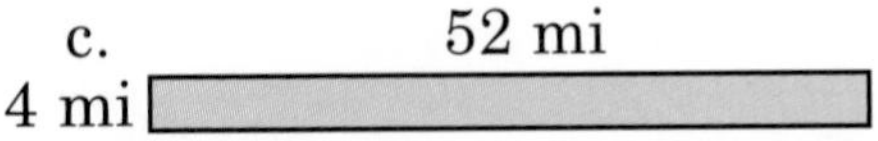


What's the area of the rectangle?

## Part 9

a. A board is 9 yards long. How many feet long is the board?

b. A tank has 100 quarts of water in it. How many gallons are in the tank?

## Part 10

a. Jim had 245 baseball cards. He bought 25 more baseball cards. His sister gave him 40 baseball cards. He gave some baseball cards to his friend. He ended up with 168 cards. How many cards did Jim give away?

b. A bike shop had some bikes. On Monday, the shop sold 18 fewer bikes than it sold on Tuesday. The shop sold 86 bikes on Monday. How many bikes did the shop sell on Tuesday?

# Lesson 90

## Part 1

a. .09 b. 3.04 c. .50 d. 7.02

## Part 2

- You know how to find the perimeter of a figure. You add up the length of each side.
- Some problems tell about the perimeter and ask about the length of a side.
- Here's a sample problem:

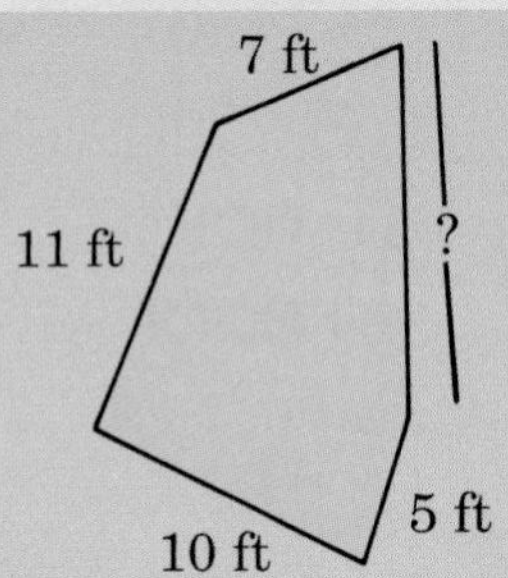


perimeter = 45 ft

- Here are the names for the number family:

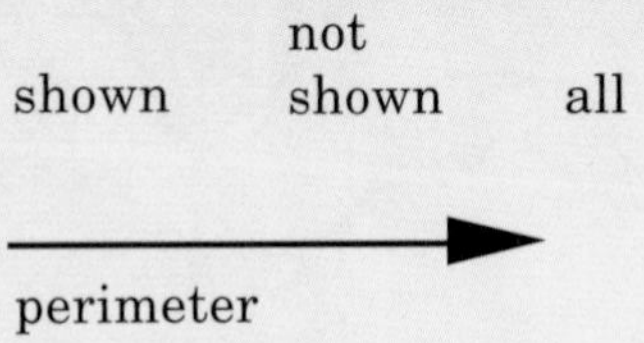


- The perimeter is the big number. You make a stack to figure out the total for the sides that are shown.
- Then you subtract that number from the big number to find the length of the missing side.

## Part 3

a.

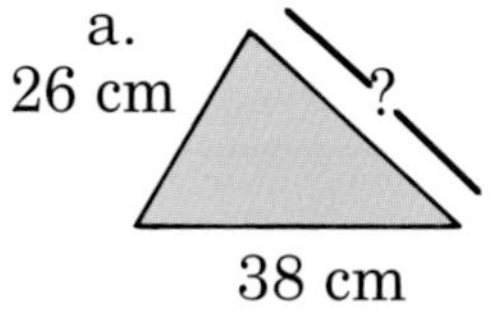


The perimeter of the figure is 96 cm. What's the length of the missing side?

b.

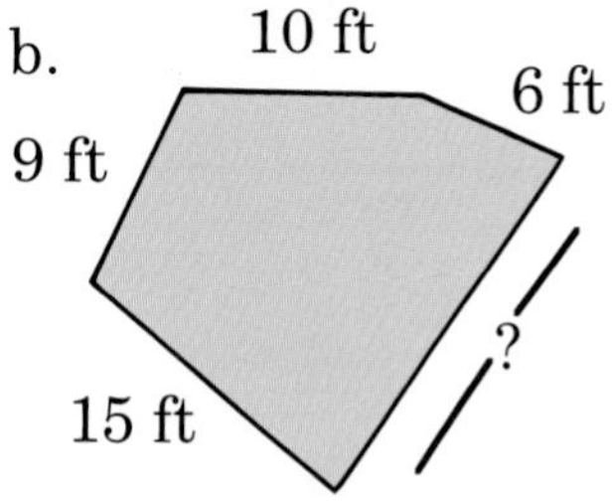


The perimeter of the figure is 56 ft. What's the length of the missing side?

c.

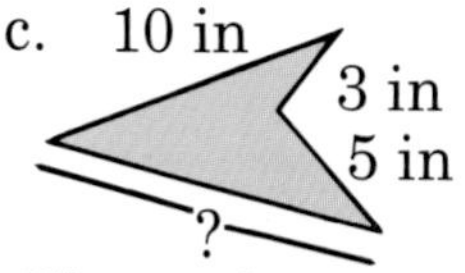


The perimeter of the figure is 29 in. What's the length of the missing side?

# Lesson 91

## Part 1

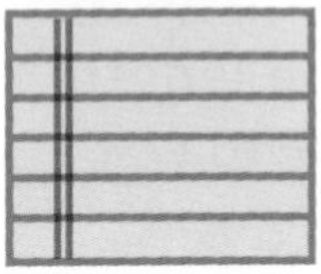

a. .08 b. 7.60 c. .80 d. 5.03

## Part 2

- Some fractions tell about what you expect to happen. These fractions don't tell what **will** happen, but what **probably** will happen.
- Here's a bag with X's and O's in it:

- There are more X's than O's. So if all the things in the bag were mixed up, and you reached into the bag without looking and pulled out one thing, it would probably be an X. Maybe it wouldn't be an X. But a smart bet would be that you would pull out an X.
- Here's a different bag with X's and O's:

- Here are two bags with X's and O's in them:

a.

$\frac{2}{8}$

b.

$\frac{2}{5}$

- You can use fractions to tell about your chances of pulling out an X from a bag. For that fraction, **the denominator tells the number of things that are in the bag.**
- **The numerator tells the number of X's.**
- For bag B, there are 5 things. There are 2 X's.
- So the fraction is $\frac{2}{5}$.
- **The closer the fraction is to 1, the better your chances are of pulling out an X.**

## Part 3

a.

b.

c.

d.

e. Which fraction gives you the best chance of drawing an X?

f. Which fraction give you the second-best chance of drawing an X?

g. Which fraction gives you the worst chance of drawing an X?

## Part 4

a.
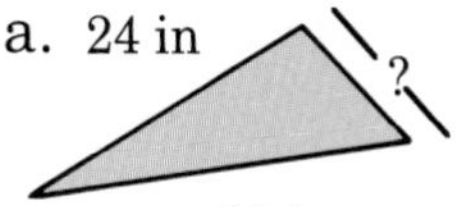


The perimeter of figure A is 67 inches. What is the length of the missing side?

b.
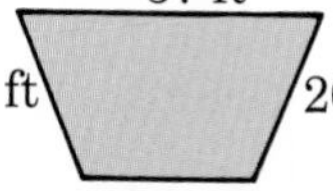


The perimeter of figure B is 100 feet. What is the length of the missing side?

c.
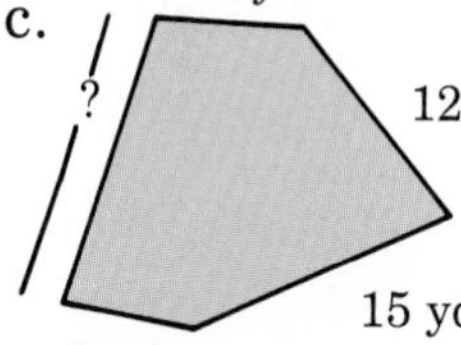


The perimeter of figure C is 57 yards. What is the length of the missing side?

## Part 5

- You can write mixed numbers for decimal numbers.
- When you read the decimal number, you say **and** for the decimal point.
- If you say **plus** for the decimal point, you say the mixed number.
- Here's 6 and 4-hundredths: **6.04**
  That's 6 plus 4-hundredths.
- Here's the mixed number for 6 and 4-hundredths: $\mathbf{6 + \frac{4}{100}}$
- Remember, if you say **plus** for the decimal point, you'll say the mixed number you write.

## Part 6

- Here's a thousands numeral: **6543**
- You can round the numeral to tens, hundreds or thousands.
- When you round it to tens, the digit **after** the tens digit is zero.
- When you round it to hundreds, the digits after the hundreds digit are zeros.
- When you round it to thousands, the digits after the thousands digit are zeros.

To round the numeral to thousands, you follow these rules:

- Make a line after the thousands digit.
- If the first digit after that line is 5 or more, round the thousands digit **up.** Then write three zeros.
- If the first digit after the line is less than 5, don't round up. Just change the digits after the line into zeros.

## Independent Work

## Part 7

**For each problem, find the fact in the table of measurement facts that tells about both names. Write the names and the ratio equation and figure out the answer.**

a. Jim bought 4 kilograms of grass seed. How many grams of grass seed did Jim buy?

b. A board is 102 feet long. How many yards long is it?

## Part 8

**Copy each problem and complete the equation.**

a. 39 = 6 x ■ + R ■
■

b. 56 = 6 x ■ + R ■
■

c. 51 = 6 x ■ + R ■
■

d. 27 = 4 x ■ + R ■
■

e. 50 = 9 x ■ + R ■
■

f. 6 = 9 x ■ + R ■
■

## Part 9

**Copy each item. Write the equation to show the fraction each decimal number equals.**

a. 7.04  b. .86  c. .02  d. 30.14

## Part 10

**Make a number family for each problem and box the answer.**

a. A deck of cards has 52 cards. 13 of those cards are diamonds. What fraction of the deck is not diamonds?

b. Sam and Jill played a game with a deck of cards. Sam had 42 cards. Jill had the rest of the cards in the deck. How many cards did Jill have?

c. There were cars on a lot. $\frac{17}{42}$ of the cars were dirty. The rest were clean. What is the fraction for all the cars on the lot? How many cars were clean?

## Part 11

**For each row, write the multiplication equation, the division problem and the answer and the fraction equation.**

| | □ x __ = □ | ⟌ | $\frac{□}{□}$ = |
|---|---|---|---|
| a. | 4 x __ = 36 | | |
| b. | | | $\frac{8}{8}$ |
| c. | | $9\overline{)0}$ | |

## Part 12

**Write a ratio equation for each problem and box the answer.**

a. Sally earned $9 every 2 hours. If she earned $288, how many hours did she work?

b. The Smith farm grows 7 bushels of beans for every 5 bushels of corn. The Smith farm grew 350 bushels of corn. How many bushels of beans did the farm grow?

## Part 13

**Figure out the answer for each item.**

a.

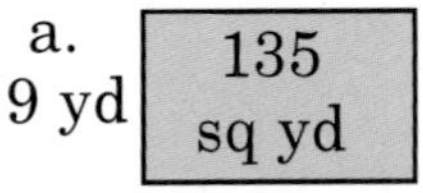


What is the width of the rectangle?

b.

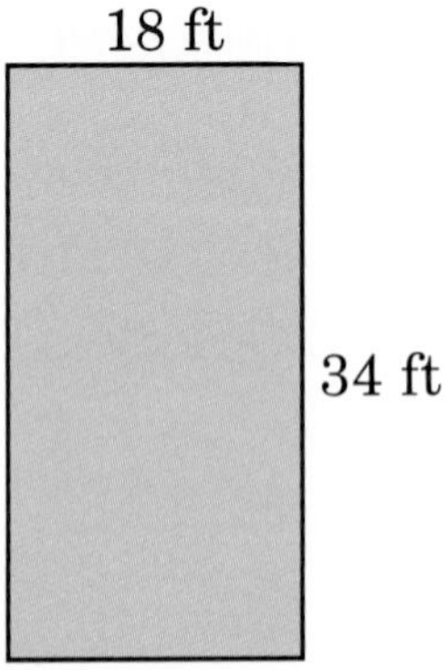


What is the area of the rectangle?

c.

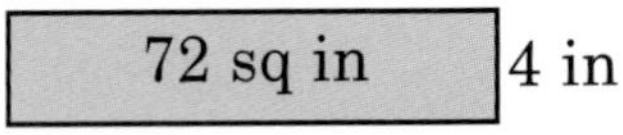


What is the width of the rectangle?

**Part J**

b.

| shown | not shown | all |
|---|---|---|
| 20 | | |
| 37 | | |
| + 20 | | |
| 77 | → perimeter | 100 |

c.

| shown | not shown | all |
|---|---|---|
| 8 | | |
| 12 | | |
| 15 | | |
| + 7 | | |
| 42 | → perimeter | 57 |

**Part K**

a. $\begin{array}{r} \overset{1}{\$3.62} \\ \times\ 12 \\ \hline 7\ 24 \\ +\ 36\ 20 \\ \hline \$\ 43.44 \end{array}$

b. $\begin{array}{r} \$7.04 \\ \times\ 9 \\ \hline \$\ 63.36 \end{array}$

c. $\begin{array}{r} \$1.\overset{7}{8}\overset{2}{3} \\ \times\ 90 \\ \hline \$164.70 \end{array}$

# Lesson 92

## Part 1

- Numbers with decimal points are different from numbers without decimal points.
- Here's a number: **5**
- If you add a zero after the 5, the number becomes much larger than 5: **50**
- When you add a zero after a decimal point, you don't change the value.
- Here's 5 with a decimal point but no zeros: **5.**
- Here's 5 with zeros after the decimal point: **5.00**
- Both values are still 5. They both tell about 5 whole units.
- The same thing is true of dollars.
- Here's 5 dollars: **$5.00**
- This is not 500 dollars. It's just 5 dollars and no cents. The zeros after the decimal point do not make the amount more than 5 dollars.

3.00 400 9 100 5 2.00

## Part 2

a.

b.

c.

d.

e. What is the best chance of drawing an X?

f. What is the second-best chance of drawing an X?

g. What is the worst chance of drawing an X?

## Part 3

a. 3.04 b. 13.09 c. 100.40 d. 22.79

## Part 4

a. $6 + \frac{2}{100}$ b. $15 + \frac{20}{100}$ c. $1 + \frac{30}{100}$ d. $200 + \frac{6}{100}$ e. $3 + \frac{99}{100}$

## Part 5

a.

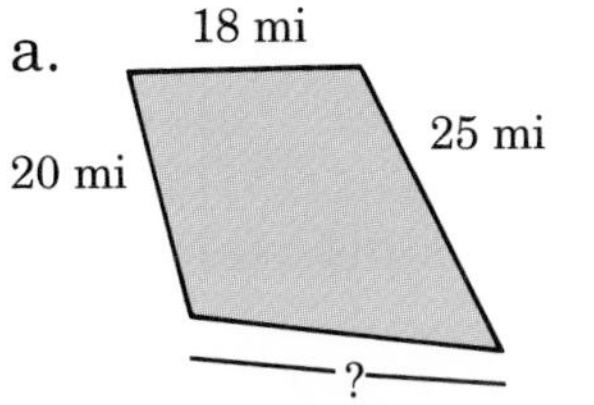


The perimeter of the figure is 89 miles. What is the length of the missing side?

b.

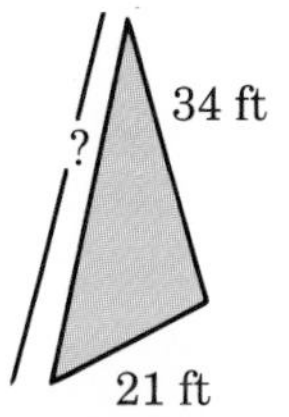


The perimeter of the figure is 99 feet. What is the length of the missing side?

c.

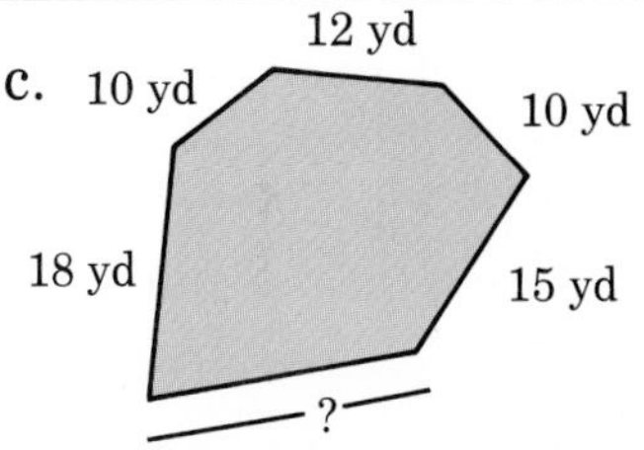


The perimeter of the figure is 87 yards. What is the length of the missing side?

## Part 6

a. Each broom costs \$3.12. Tim wants to buy 4 brooms. How much would he spend?

b. Each cup costs \$1.04. Donna wants to buy 12 cups. How much would she spend?

c. Each post costs \$3.60. Fran wants to buy 9 posts. How much money will she need?

## Independent Work

## Part 7

**Write a ratio equation for each problem and box the answer.**

a. If 3 boys can pull 2 logs, how many boys are needed to pull 192 logs?

b. The ratio of bass to all the fish in a lake is 3 to 8. There are 540 bass in the lake. How many total fish are in the lake?

c. If 6 boards weigh 5 pounds, how much do 600 boards weigh?

## Part 8

**Copy the problems and write the answers.**

| a. | b. | c. | d. | e. | f. | g. |
|---|---|---|---|---|---|---|
| 8 | 8 | 8 | 8 | 8 | 8 | 8 |
| x 7 | x 3 | x 9 | x 4 | x 8 | x 0 | x 6 |

## Part 9

**Make a number family for each problem. Figure out the answer.**

a. A bus had 20 people on it. At the first stop, 3 people got off the bus and 2 got on the bus. At the next stop, 11 people got off the bus and 1 person got on the bus. At the next stop, 4 people got off the bus. How many people were still on the bus?

b. A tank has some water in it. In the morning, 145 gallons were removed from the tank. In the afternoon, 38 gallons were removed. There were still 356 gallons of water in the tank. How many gallons did the tank start out with?

c. 19 trees in the forest were dead and the rest were living. There were 189 total trees in the forest. What fraction of the trees were living?

d. A tank contained 860 gallons of water. A tub contained 538 gallons of water. How many fewer gallons were in the tub than in the tank?

## Part 10

**Copy each item. Write the equation to show the fraction each decimal number equals.**

a. 10.07

b. 4.60

c. .05

d. 7.08

## Part 11

**Copy the table and figure out the missing numbers.**

| | | | Total |
|---|---|---|---|
| | 39 | | 620 |
| | | | |
| Total | 158 | | 949 |

# Lesson 93

## Part 1

a. 

b. 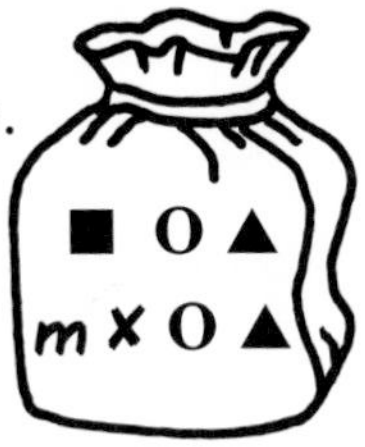

c. 

d. 

## Part 2

***Menu***

| | |
|---|---|
| Yogurt | $ .65 |
| Soup | $1.11 |
| Hamburger | $1.45 |
| Fish Sandwich | $2.15 |

a. Tim wants to buy a hamburger and soup. How much money does he need?

b. Jan orders 6 hamburgers for the people at her table. How much money does Jan need?

c. Jack orders a fish sandwich and yogurt. How much money does he spend?

d. Ginger had $7.20. She bought a fish sandwich. How much money does she have left?

e. Slim ordered 11 fish sandwiches. How much did Slim spend?

## Part 3

1.00    10    70    13.00    7.00    2    .12

## Part 4

a.

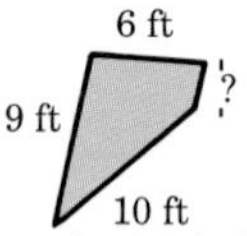


The perimeter is 27 feet. What is the length of the missing side?

b.

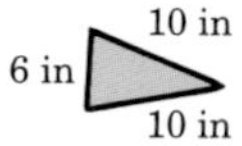


What is the perimeter of the triangle?

c.

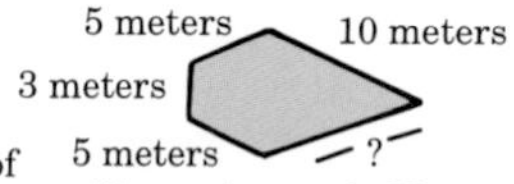


The perimeter is 33 meters. What is the length of the missing side?

d.

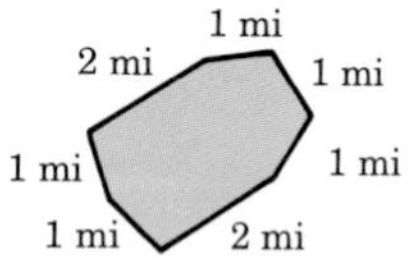


What is the perimeter of the figure?

## Part 5

- Here's how to write a fraction that tells about your chances of drawing a square from a bag:

  **Write the number of squares in the numerator and the total number of things in the denominator.**

- Here's a bag with 6 things:
  The fraction is $\frac{1}{6}$:

**$\frac{1}{6}$**

- You can tell about your chances of getting a square by calling the denominator of the fraction **trials** and calling the numerator **expected winners.**

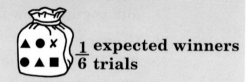

**1 expected winners**
**6 trials**

- The number of **trials** is the number of times you reach into the bag and pull out something. The number of **expected winners** is the number of squares you expect to get from those trials. You put the thing back in the bag after each trial.
- For $\frac{1}{6}$, you could say: **If I took 6 trials, I would expect to get 1 winner.**
- For this fraction, you could say: If I took 8 trials, I would expect to get 3 winners. **$\frac{3}{8}$**
- Here's another fraction: **$\frac{5}{7}$**
- Here's another fraction: **$\frac{3}{10}$**

## Independent Work

### Part 6 Copy each problem and complete the equation.

a. $5 = 7 \times \blacksquare + R\ \blacksquare$
$\blacksquare$

b. $11 = 5 \times \blacksquare + R\ \blacksquare$
$\blacksquare$

c. $26 = 7 \times \blacksquare + R\ \blacksquare$
$\blacksquare$

d. $38 = 4 \times \blacksquare + R\ \blacksquare$
$\blacksquare$

### Part 7 Copy each problem and work it.

a. $3 \times \frac{4}{5} =$

b. $3 + \frac{4}{5} =$

c. $\frac{12}{3} - 1 =$

d. $\frac{3}{4} \times \frac{5}{5} =$

e. $\frac{7}{9} - \frac{2}{9} =$

### Part 8 Write a ratio equation for each problem and box the answer.

a. Sam had goldfish and tadpoles. He had 3 goldfish for every 7 tadpoles. If he had 630 tadpoles, how many goldfish did he have?

b. There were 3 girls in the line for every 4 boys in the line. If there were 60 boys in line, how many girls were there?

### Part 9 Copy each equation and complete it.

a. $4 = \frac{\phantom{0}}{6}$

b. $\frac{36}{9} = \blacksquare$

c. $\frac{\phantom{0}}{5} = 9$

d. $\blacksquare = \frac{24}{6}$

## Part 10

**Write more than 1, less than 1, or = 1 to tell about each missing value.**

a. $3 \times ■ = \frac{1}{5}$

b. $4 \times ■ = \frac{20}{5}$

c. $4 \times ■ = 8$

d. $4 \times ■ = \frac{5}{8}$

## Part 11

**Copy the table. Put in all the missing numbers. Then answer the questions.**

This table is supposed to show the number of trout and other fish caught at two places on a river.

| | Big Bend | Deep Hole | Total for both places |
|---|---|---|---|
| Trout | 14 | | 52 |
| Other fish | | | |
| Total fish | 60 | 81 | |

***Questions***

a. Were there more fish caught at Big Bend or at Deep Hole?

b. How many trout were caught?

c. How many total fish were caught?

d. At Pike Point, there were 15 more fish caught than there were at Big Bend. How many fish were caught at Pike Point?

e. At Wet Rock, there were 11 fewer fish caught than there were at Big Bend. How many fish were caught at Wet Rock?

## Part 12

**Copy the problems and figure out the answers.**

a. $6\overline{)9612}$

b. $3\overline{)2916}$

## Part J

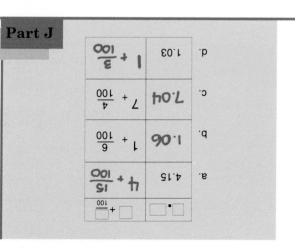

| | | |
|---|---|---|
| | □.□ | $□ + \frac{□}{100}$ |
| a. | 4.15 | $4 + \frac{15}{100}$ |
| b. | 1.06 | $1 + \frac{6}{100}$ |
| c. | 7.04 | $7 + \frac{4}{100}$ |
| d. | 1.03 | $1 + \frac{3}{100}$ |

# Lesson 94

**Part 1**

a. Mrs. Jones bought 4 cassette tapes. How much did she spend?

b. Mrs. Smith had $36.00. She bought the radio. How much money did she have left?

c. Frank needs $12.50 to buy the radio. How much money does Frank have?

d. Ginger buys 3 compact discs. How much does she spend?

e. Dan buys a cassette tape, a compact disc and a radio. How much does he spend?

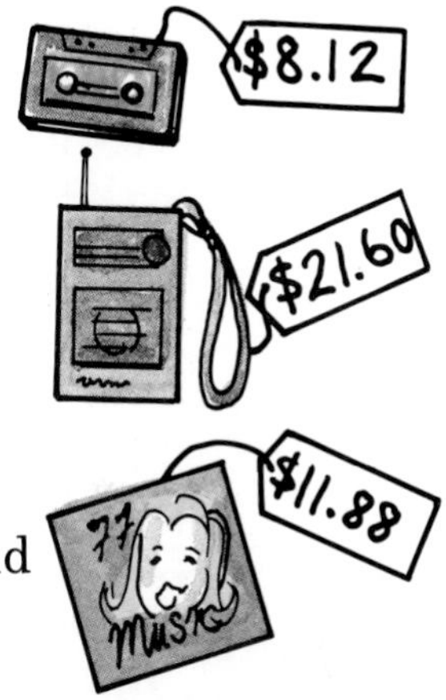


**Part 2**

- If whole numbers do not have a decimal point, you can make a decimal point after the last digit. The decimal point does not change the value.
- Here's 7: **7**
- Here's 7 with a decimal point: **7.**
- Here's 7 with a decimal point and zeros after the decimal: **7.00**
- If you make a decimal point after the last digit of whole numbers, you can line up the decimal points and see which numbers are biggest.

| Here are some numbers. Some don't have decimal points. | Here they are with decimal points. The decimal points are lined up. |
|---|---|
| **15** | **15.** |
| **2.00** | **2.00** |
| **1.50** | **1.50** |
| **7** | **7.** |
| **.07** | **.07** |

10 3.00 11.00 8.04 5 .27 8.84

## Part 3

a. The grill can cook 6 hamburgers at a time. It takes 8 minutes to cook 6 hamburgers. How long will it take to cook 30 hamburgers?

b. A repair shop can paint 5 cars in 3 hours. If the shop paints cars for 27 hours, how many cars does it paint?

c. Every 7 seconds, a fish moves 9 feet. How long does it take the fish to move 540 feet?

# Independent Work

## Part 4

**Figure out the area or the length of the missing side. Then figure out the perimeter of each rectangle.**

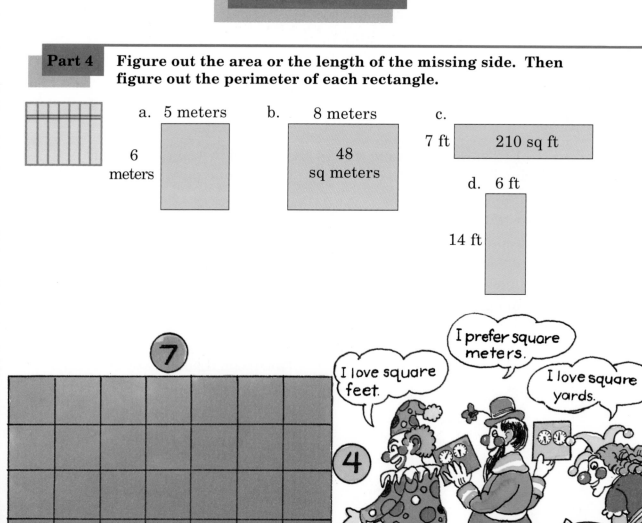

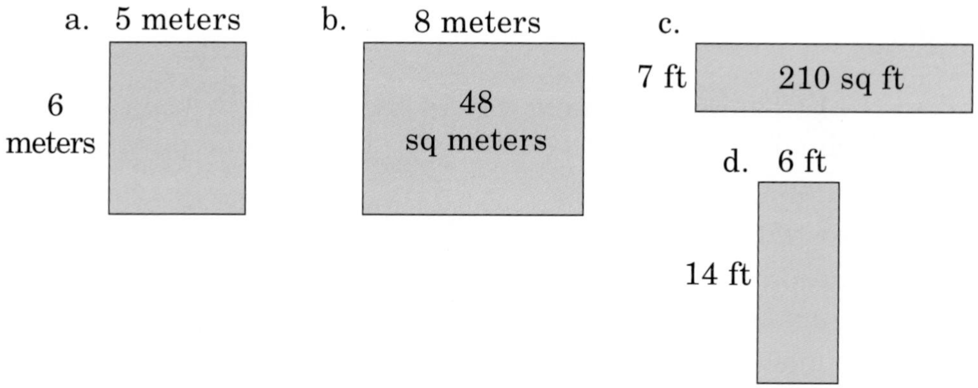

## Part 5 Make a number family for each problem and box the answer.

a. Tom's mother is 34 years older than Tom. Tom's mother is 49. How old is Tom?

b. Mrs. Jones sells flowers. On Monday, she started out with 119 flowers. She sold 16 in the morning. She sold 18 in the afternoon. Then she bought 50 flowers from her neighbor. How many flowers did she end up with?

c. In Fern Park there were 34 pine trees. The rest were fir trees. There were 88 trees in all. How many were fir trees? What **fraction** of the trees were pine trees? What's the **fraction** for all the trees?

d. Tina had $13.20. She bought lunch for $1.85. She bought pencils for $1.11. Tim gave Tina $21.00 for her radio. How much money did Tina end up with?

## Part 6 Copy each equation and complete it.

a. $4 = \frac{\quad}{24} = \frac{\quad}{3} = \frac{\quad}{7} = \frac{\quad}{5}$

b. $17 = \frac{\quad}{3} = \frac{\quad}{2} = \frac{\quad}{5}$

## Part 7 Copy each problem and work it.

a.
$$\begin{array}{r} \$3.20 \\ \underline{\times \quad 7} \end{array}$$

b.
$$\begin{array}{r} \$41.12 \\ \underline{\times \quad 53} \end{array}$$

c.
$$\begin{array}{r} \$3.02 \\ \underline{\times \quad 2} \end{array}$$

# Lesson 95

## Part 1

**_Sample Problem:_** 3 pounds of grass seed cost $4.30. How much do 15 pounds cost?

a. 7 bulbs cost $1.06. How much do 28 bulbs cost?

b. 3 pints of plant food cost $1.30. How much do 21 pints cost?

## Part 2

a. **2**301  b. **6**501  c. 69**5**3  d. 1**4**20  e. 10**3**9  f. 95**8**2

## Part 3

1.90  9.03  850  .86  .05  6  1

## Part 4

- You can make number families that show the number of winners and losers in a set.
- Here's a set:

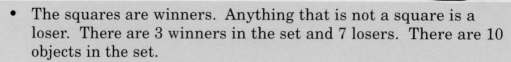

- The squares are winners. Anything that is not a square is a loser. There are 3 winners in the set and 7 losers. There are 10 objects in the set.
- So here is the number family for the winners and losers:

**winners** **losers** **all**
**3** **7** → **10**
**objects**

- You can write **fractions** for the winners and losers in the set. The whole set is 1 whole. So the fraction for the total is $\frac{10}{10}$.

- Here's the number family with the fractions:

| winners | losers | all |
|---|---|---|
| $\frac{3}{10}$ | $\frac{7}{10}$ | → $\frac{10}{10}$ |
| **objects** | | |

- The fractions tell you that, if you took 10 trials, you'd expect to get 3 winners and 7 losers.

- Here's a problem:

  **The winners for this problem are aces. There are 52 cards in a deck of cards. There are 4 aces.**

- Here's the number family that shows the number of winners and the number of cards.

| aces | not aces | all |
|---|---|---|
| 4 | | → 52 |
| **cards** | | |

- The name for winners is always the first small number in the family.

| aces | not aces | all |
|---|---|---|
| 4 | 48 | → 52 |
| **cards** | | |

- Here are the fractions that tell about the winners and losers.

| aces | not aces | all |
|---|---|---|
| $\frac{4}{52}$ | $\frac{48}{52}$ | → $\frac{52}{52}$ |
| **cards** | | |

- All the cards in the deck make up 1 whole. So the fraction for **all** is $\frac{52}{52}$.

**Part 5**

a. The winners for this set are candies with nuts. There are 40 pieces of candy in the box. 23 pieces of candy have nuts.

b. Winners for this set are red stones. There are 70 stones in a bag. 29 stones are red.

## Part 6

- Here's a division sign that looks like a fraction: $\div$
- It's a division sign.
- The dots show where the numerator and denominator go. The bar is like the bar of a fraction.
- When you read a division problem that has this sign, you say **divided by** for the sign. What you say tells you how to write the fraction.
- Here's 32 divided by 4: $\mathbf{32 \div 4 =}$
- Here's the fraction: $\mathbf{32 \div 4 = \frac{32}{4}}$
- Here's 63 divided by 7: $\mathbf{63 \div 7 =}$
- Here's the fraction: $\mathbf{63 \div 7 = \frac{63}{7}}$

a. $24 \div 4 =$ b. $48 \div 6 =$ c. $72 \div 9 =$ d. $24 \div 3 =$

## Independent Work

## Part 7

**For each item, write the fractions for both pictures. Complete the equation to show the fraction of 1 you multiply the first fraction by to get the second fraction.**

a.
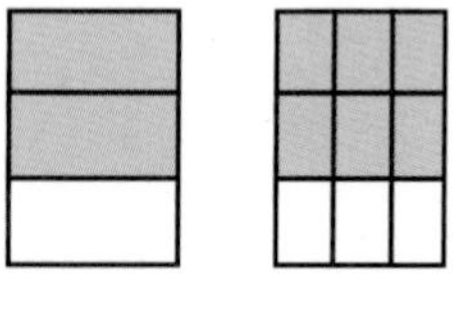

b.
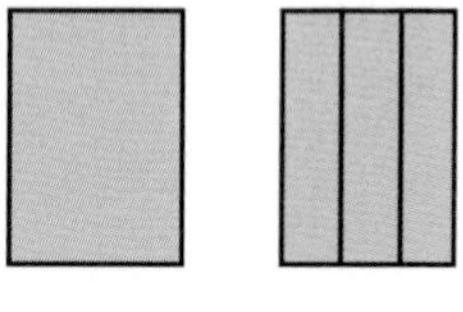

c.
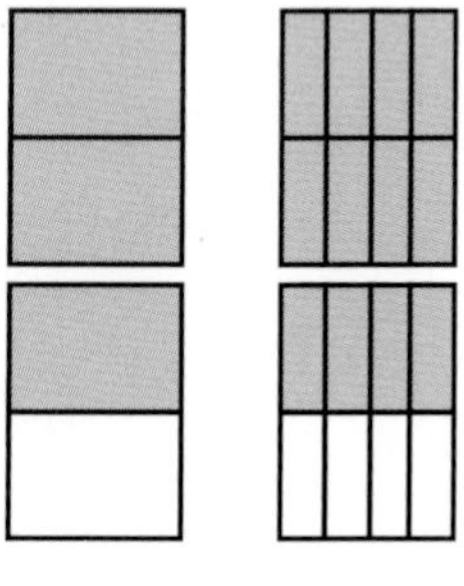

## Part 8 Copy each problem and complete the equation.

a. 17 = 5 x ■ + R ■
   ■

b. 50 = 7 x ■ + R ■
   ■

c. 44 = 6 x ■ + R ■
   ■

d. 27 = 4 x ■ + R ■
   ■

## Part 9 Make a number family for each problem and box the answer.

a. A truck was empty. Then 650 pounds of sand were put in the truck. Then another 1160 pounds were put in. Then some pounds were removed. Then 345 pounds were put in. 1438 pounds of sand were left in the truck. How many pounds of sand were removed?

b. There are 24 blue cards in a pack. The rest of the cards are not blue. There are 91 cards in the pack. How many cards are not blue? What fraction of the cards are blue?

c. Mary had some money. She spent $4.64 at Z Mart. She spent $12.88 at R Mart. Then she spent $40.12 at the Bosh Mart. She ended up with $45.01. How much money did she start out with?

d. Swan Lake is 123 feet deeper than Fern Lake. Swan Lake is 1891 feet deep. How deep is Fern Lake?

e. A motorcycle goes 37 miles per hour slower than a race car. The motorcycle goes 181 miles per hour. How fast does the race car go?

Let's see. I'm going 181 miles per hour.
That's nothing.

# Lesson 96

## Part 1

a. A garden is in the shape of a rectangle. The area of the garden is 140 square feet. The garden is 7 feet wide. How long is the garden?

b. A floor is in the shape of a rectangle. The floor has an area of 234 square feet. The floor is 9 feet wide. How long is the floor?

c. A rectangular room was 12 feet wide and 16 feet long. What was the area of the room?

d. A cardboard top is 4 inches wide. The area of the cardboard top is 128 square inches. How long is the cardboard top?

## Part 2

a. $56 \div 7 =$  b. $12 \div 2 =$  c. $30 \div 6 =$

## Part 3

a. 5 pounds of flour cost $1.05. How much do 35 pounds of flour cost?

b. 3 ears of corn cost $.89. How much do 18 ears of corn cost?

c. 7 ounces of salad dressing cost $2.18. How much do 28 ounces cost?

## Part 4

a. The winners for this set are trout. There are 13 trout in a lake. There are 15 other fish in the lake.

b. The winners for this set are clear marbles. There are 78 marbles in a bag. 49 of the marbles in the bag are clear.

c. The winners for this set are cards that have pictures on them. There are 52 cards in a deck. There are 12 cards in a deck that have pictures on them.

## Independent Work

**Part 5** **Work the problems. For some problems, write a ratio equation. For the other problems, make a number family.**

a. The ratio of yards to rods is 11 to 2. If something is 40 rods long, how many yards long is it?

b. Tina is 13 years younger than her brother. Her brother is 39 years old. How old is Tina?

c. A bus started out with no passengers on it. At the first stop, 11 passengers got on. At the next stop, 12 passengers got on, and 2 passengers got off. How many passengers ended up on the bus?

d. The distance from Chicago to New York is 714 miles. The distance from Chicago to Denver is 920 miles. How much farther is the trip to Denver than the trip to New York?

e. Hill Town has 378 more houses than Valley Town. There are 900 houses in Valley Town. How many houses are in Hill Town?

f. 230 shoes in the shoe store were white. The rest were not white. If there were 998 shoes in all, how many were not white? What is the fraction for white shoes?

**Part 6** **Copy the table and write the $x$ and $y$ values for A, B and C.**

| | $x$ | $y$ |
|---|---|---|
| A | | |
| B | | |
| C | | |

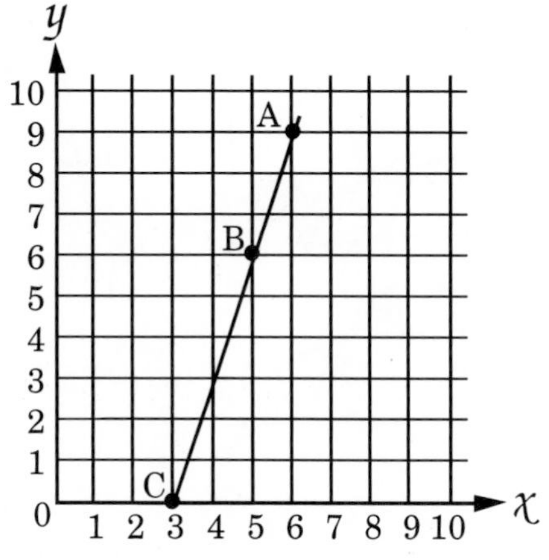

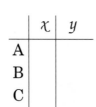

**Part 7** Order the numbers. Write the largest number first and the smallest number last. Show a decimal point in each number.

7.03 196 19.90 .85 16 .09

**Part J**

| | Decimal Number | Mixed Number | Fraction |
|---|---|---|---|
| c. | 1.67 | $1 + \frac{67}{100}$ | $\frac{167}{100}$ |
| d. | 10.50 | $10 + \frac{50}{100}$ | $\frac{1050}{100}$ |
| e. | 71.09 | $71 + \frac{9}{100}$ | $\frac{7109}{100}$ |

# Lesson 97

## Part 1

**Nuts**

| | |
|---|---|
| 2 pounds of peanuts cost | $1.75 |
| 2 pounds of cashews cost | $5 |
| 2 pounds of hickory nuts cost | $2.50 |
| 2 pounds of walnuts cost | $3.02 |

a. How much do 10 pounds of hickory nuts cost?

b. How much do 6 pounds of walnuts cost?

c. Fran spent $35 for cashews. How many pounds did she buy?

## Part 2

a. The winners for this set are trout. There are 132 trout in a lake. There are 156 other fish in the lake.

b. The winners for this set are seeds that sprouted. 470 seeds were in a bag. 320 of the seeds sprouted. The rest did not sprout.

c. The winners for this set are coins that are silver. Jane had a piggy bank with 4 silver coins and 39 coins that were not silver.

## Part 3

- For some problems, you have to add a lot of hundreds numbers or a lot of thousands numbers.
- You can work the problem with rounded numbers to get an answer close to the actual answer.
- Here's a problem:

$$\begin{array}{r} 291 \\ 531 \\ \underline{+\ 468} \end{array}$$

- All are hundreds numbers. So you round to hundreds.

**291 rounds to 300**
**531 rounds to 500**
**468 rounds to + 500**

- You add the **first digit** of the rounded numbers:
- So the answer is 13 hundred. That's close to the answer you'd get if you added the actual numbers.

**hundreds**

$$\begin{array}{r} 3 \\ 5 \\ \underline{+\ 5} \\ 13 \end{array}$$

a.
$$\begin{array}{r} 8824 \\ 2387 \\ 1072 \\ \underline{+\ 5118} \end{array}$$

b.
$$\begin{array}{r} 179 \\ 312 \\ 427 \\ \underline{+\ 858} \end{array}$$

c.
$$\begin{array}{r} 3467 \\ 7598 \\ \underline{+\ 1194} \end{array}$$

## Part 4

a. The area of a rectangular piece of glass was 918 square inches. The glass was 9 inches wide. How long was the piece of glass?

b. A ranch was in the shape of a rectangle. The area of the ranch was 54 square miles. The ranch was 6 miles wide. How long was the ranch?

c. A rectangular garden is 37 yards long and 45 yards wide. What is the area of the garden?

d. A room is 7 feet wide and has an area of 287 square feet. What is the length of the room?

## Part 5

- If you start with a coordinate system that has points shown in a line, you can figure out the function. We know that the function rule for the $x$ and $y$ values is the same because the points are in a straight line.
- Here's a coordinate system with points on a line:
- To figure out the function, first write the $x$ value and $y$ value for each point.

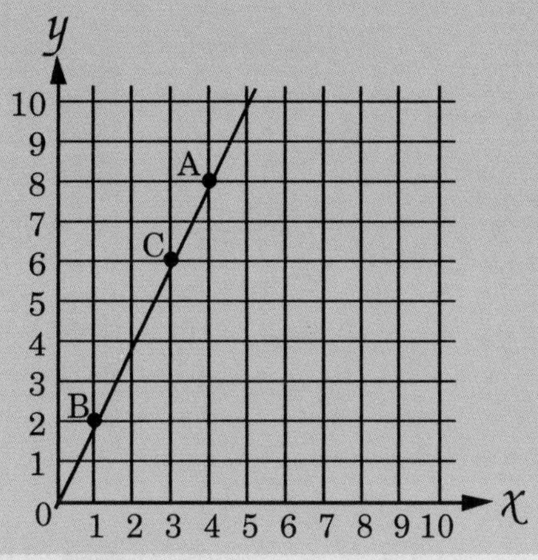


- Here's the $x$ and $y$ values in the function table:
- Now you can figure out the function rule that starts with $x$.

| | | Function | Answer |
|---|---|---|---|
| | $x$ | | $y$ |
| A | 4 | | 8 |
| B | 1 | | 2 |
| C | 3 | | 6 |

- For row A, you ask: What are the ways of getting from 4 to 8?
- Then you look at the next row and see which of those function rules work.

| | | Function | Answer |
|---|---|---|---|
| | $x$ | $x + 4$<br>$x \times 2$ | $y$ |
| A | 4 | | 8 |
| B | 1 | | 2 |
| C | 3 | | 6 |

- You ask: Which function rule works for $x = 1$ and $y = 2$?
- $x \times 2$ is the function for all of the points on the line.

| | | Function | Answer |
|---|---|---|---|
| | $x$ | ~~$x + 4$~~<br>$x \times 2$ | $y$ |
| A | 4 | | 8 |
| B | 1 | | 2 |
| C | 3 | | 6 |

## Independent Work

**Part 6** **The winners for these sets are stars. For each set, write the fraction. Copy and complete the statement.**

a.

If I took __ trials, I would expect __ winner(s).

b.

If I took __ trials, I would expect __ winner(s).

c.

If I took __ trials, I would expect __ winner(s).

**Part 7** **Answer the question for each item.**

a.

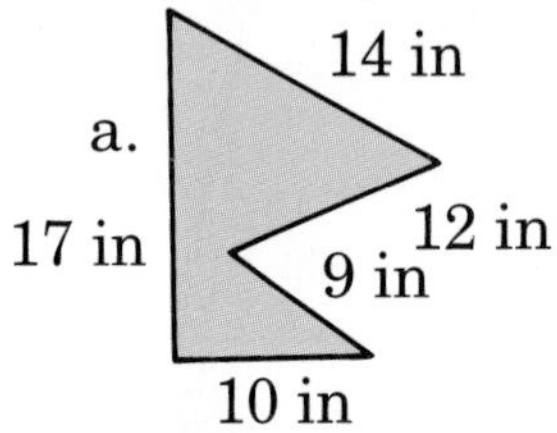


What is the perimeter?

b.

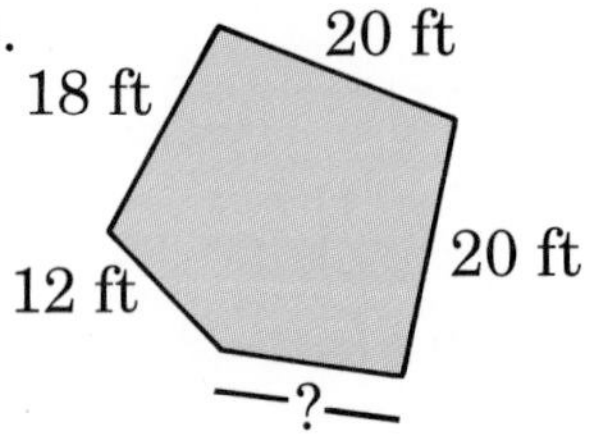


The perimeter is 84 feet. What is the length of the missing side?

c.

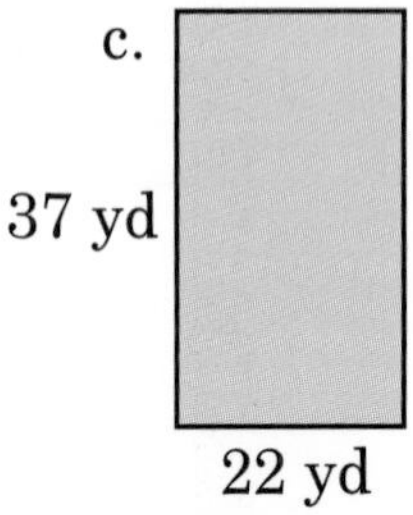


What is the perimeter?

d. What is the area of rectangle C?

**Part 8** **For each item, write the fraction.**

a.

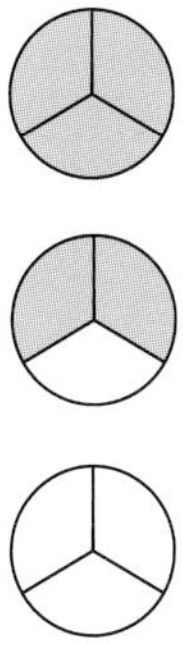

b. The numbers are 81 and 72. The fraction is less than 1.

c. The numbers are 81 and 72. The denominator is 81.

d. The picture shows 15 parts shaded. There are 107 parts in each unit.

e. The numerator is 64. The denominator is 49.

f. The numbers are 17 and 4. The numerator is 4.

# Lesson 98

## Part 1

- You've learned to write fractions for expected winners and losers. You can also work **ratio problems** for winners and trials.
- Here's a set:
  The winners are blue circles. The number of things in the set shows the number for trials.

- Here's the names and the fraction for the ratio equation: $\frac{\textbf{winners}}{\textbf{trials}}\ \frac{3}{5}$
- You're going to make an equivalent fraction. That fraction equals $\frac{3}{5}$, so it will have the same ratio of winners to trials as $\frac{3}{5}$.
- You can figure your expected winners for any number of trials.
- You can also figure the number of trials you'd probably take for any number of winners.
- If you wanted to know the number of **trials** you'd probably take to get 12 **winners,** you'd work this problem: $\frac{\textbf{winners}}{\textbf{trials}}\ \frac{3}{5} = \frac{12}{\ }$
- If you wanted to know the number of expected **winners** for 30 **trials,** you'd work this problem: $\frac{\textbf{winners}}{\textbf{trials}}\ \frac{3}{5} = \frac{\ }{30}$
- If you wanted to know the number of **trials** you'd probably have to take to get 24 **winners,** you'd work this problem: $\frac{\textbf{winners}}{\textbf{trials}}\ \frac{3}{5} = \frac{24}{\ }$

a. The winners for this set are blue circles. How many winners would you expect if you took 42 trials?

b. The winners for this set are blue triangles. About how many trials would you expect to take to get 42 winners?

## Part 2

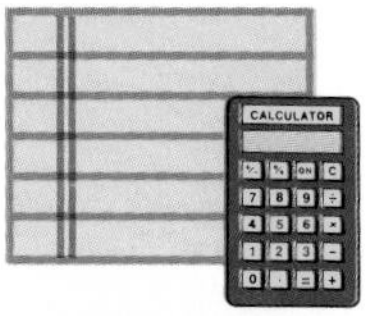

a.
$$\begin{array}{r} 2645 \\ 8291 \\ +\ 1087 \\ \hline \end{array}$$

b.
$$\begin{array}{r} 635 \\ 574 \\ +\ 185 \\ \hline \end{array}$$

c.
$$\begin{array}{r} 1461 \\ 8592 \\ +\ 9904 \\ \hline \end{array}$$

## Part 3

- These are cubes. Each side of a cube is a square.

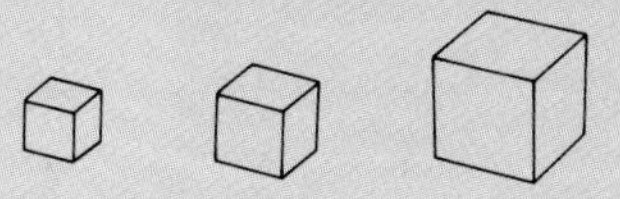

- Here's a box:
  You want to figure out the number of cubes that fit in the box. When you find the number of cubes that fit in a box, you find the **volume** of the box.

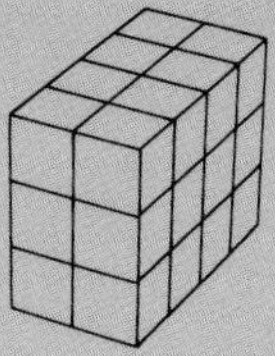

- Here's the equation for the volume of a box: **Volume = area of base x height**
- The base is the bottom side of the box: The bottom side is shaded.

- The bottom side of the box is 2 units wide and 4 units long. The area of the base is 8 squares.

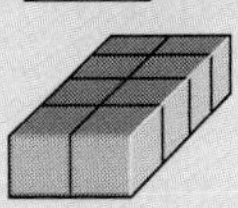

- Here's the equation with the number for the area of the base: **Volume = 8 x height**
- The height is 3 units. Here's the equation with the number for the height: **Volume = 8 x 3**
- Here's the volume: **Volume = 8 x 3 = 24 cubes**
- Here's the box split apart with the cubes shown:

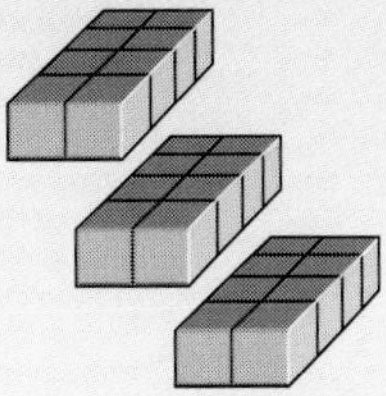

## Part 4

**Volume = area of base x height**

a.

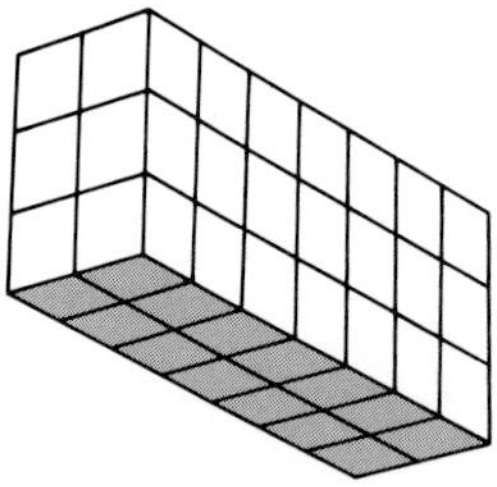

b.

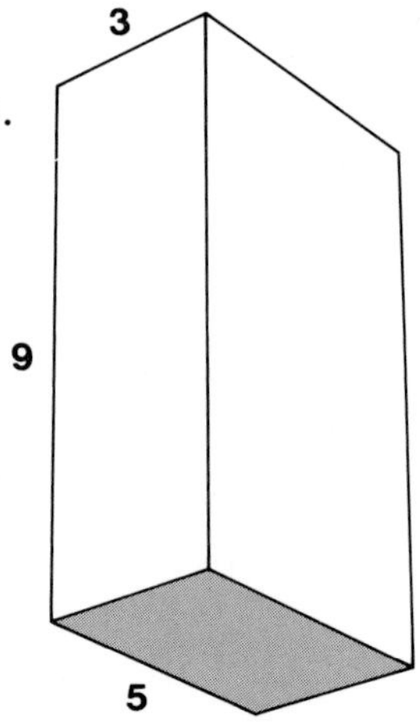


## Independent Work

## Part 5

**Make a number family or write a ratio equation for each problem. Then answer the question.**

a. Jim weighs 64 pounds more than Sara. Jim weighs 118 pounds. How much does Sara weigh?

b. 47 geese were sitting in a field. 75 geese were flying over the field. What is the fraction for geese that were sitting in the field?

c. In the marsh, there were 4 geese for every 5 ducks. 72 geese lived in the marsh. How many ducks lived in the marsh?

d. 47 geese were sitting in a marsh. 89 more geese landed in the marsh. 32 geese flew away. 8 geese walked out of the marsh. 16 geese swam out of the marsh. How many geese ended up in the marsh?

e. Jan had some money. She ate lunch at a restaurant. Her meal cost $6.90 and she tipped the waitress $1.25. On the way out, Jan bought some mints for $.65. Jan had $18.35 left. How much money did Jan have to start with?

## Part 6

**For each row, write the multiplication equation, the division problem and answer, the fraction equation and the division equation.**

| | □ x __ = □ | $\overline{)\quad}$ | $\frac{□}{□}$ = □ | □ ÷ □ = __ |
|---|---|---|---|---|
| a. | | | | 27 ÷ 9 = __ |
| b. | | $3\overline{)192}$ | | |
| c. | 5 x __ = 80 | | | |

## Part 7

**Answer the questions.**

a. Jim's room is 12 feet wide and 18 feet long. What is the area of Jim's room?

b. A rectangle has an area of 136 square inches. The rectangle is 8 inches wide. How long is the rectangle?

This rectangle is 4 feet high and it has an area of 31,444,000 square feet.
I think this rectangle is pretty wide.

# Lesson 99

## Part 1

- Some lines **intersect.** Those are lines that touch each other or cross each other.

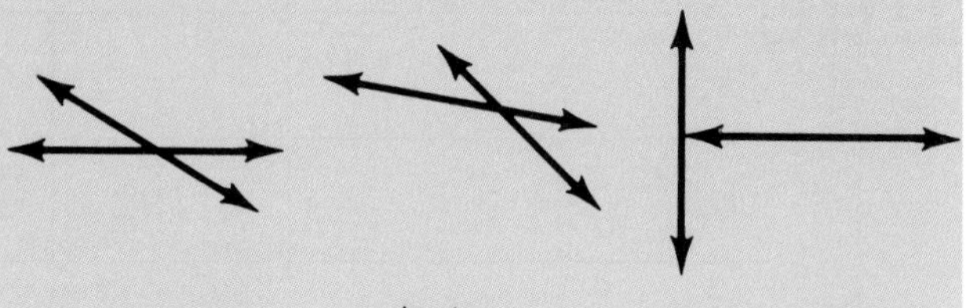

- Some lines do not intersect. Some of those lines are **parallel.** Parallel lines are always the same distance from each other.

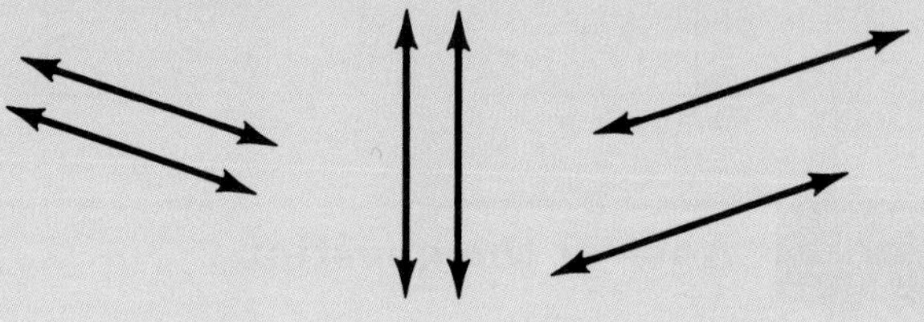

- Lines that are **not** parallel will always intersect if they go far enough.

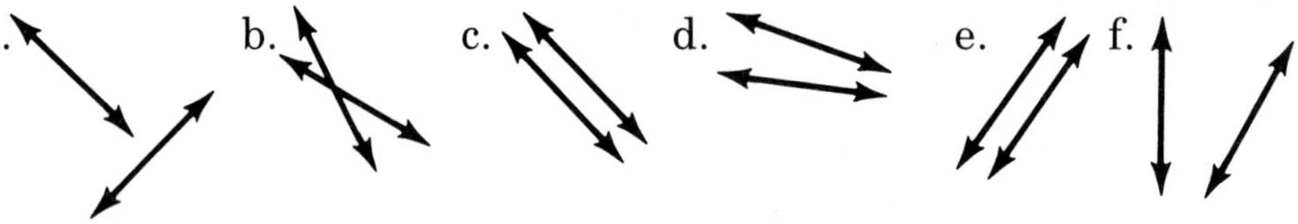


## Part 2

***Sample set:*** $\frac{8}{12}$ $\frac{8}{20}$ $\frac{8}{5}$ $\frac{8}{7}$ $\frac{8}{2}$ $\frac{8}{8}$

Set a: $\frac{5}{6}$ $\frac{5}{5}$ $\frac{5}{1}$ $\frac{5}{3}$ $\frac{5}{7}$

Set b: $\frac{10}{20}$ $\frac{10}{30}$ $\frac{10}{11}$ $\frac{10}{13}$ $\frac{10}{14}$

## Part 3

a. For this set, if you took 9 trials, you'd expect to get 2 winners. How many winners would you expect to get if you took 90 trials?

b. For this set, if you took 8 trials, you would expect to get 1 winner. About how many trials would you have to take to get 16 winners?

c. For this set, if you took 7 trials, you'd expect to get 6 winners. About how many trials would you have to take to get 36 winners?

## Part 4

**Volume = area of base x height**

a. 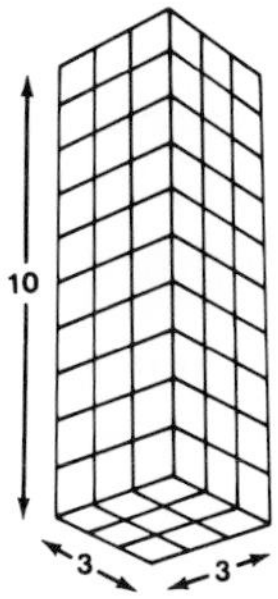


b. 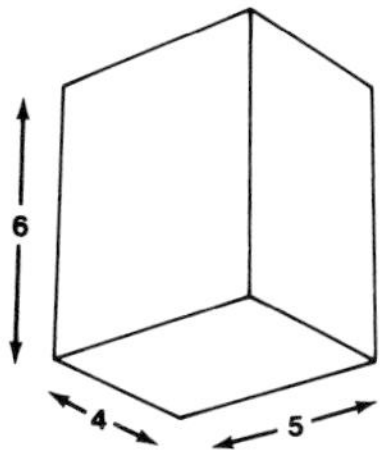


c. 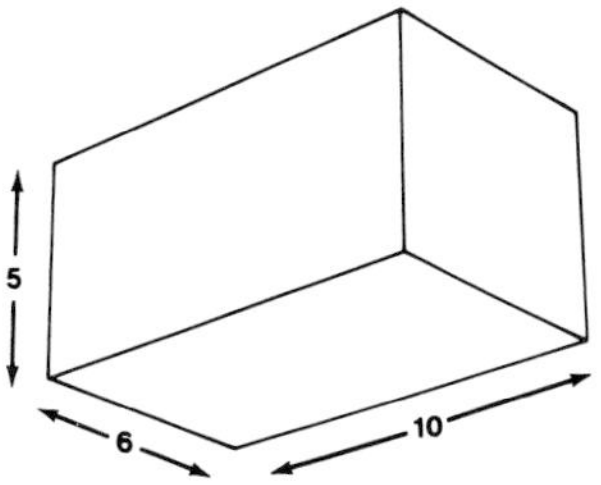


## Part 5

***Sample sentences:*** There are 5 cans in every box.
There are 5 cans per box.

a. There were 6 cans per box.

b. On the Jones' farm, there are 60 fish per pond.

c. A man ran 13 miles per hour.

d. There were 8 gallons per can.

## Independent Work

## Part 6

**For each row, write the decimal number, the fraction and the mixed number.**

| | □.□ | $\frac{□}{□}$ | □ + $\frac{□}{□}$ |
|---|---|---|---|
| a. | | | $12 + \frac{6}{100}$ |
| b. | .09 | | |
| c. | | $\frac{320}{100}$ | |

## Part 7

**For each problem, write a ratio equation or a number family. Then write the answer and box it.**

a. A farmer can plow 2 acres every 42 minutes. How long would it take a farmer to plow 84 acres?

b. A farmer planted corn and wheat. The farmer planted 42 fewer acres of corn than wheat. The farmer planted 84 acres of corn. How many acres of wheat did the farmer plant?

c. A farmer planted 183 acres. The farmer planted corn and wheat. The farmer planted 42 acres of corn. How many acres of wheat did the farmer plant?

d. For problem c, what fraction of acres did the farmer plant in corn?

e. A farmer put up a fence 6 kilometers long. How many meters long is the fence?

f. A farmer had some bushels of corn. The farmer sold 497 bushels of corn. The farmer ate 64 bushels. The farmer planted 212 bushels. The farmer ended up with 513 bushels. How many bushels of corn did the farmer have to start with?

g. For every 3 bushels of corn a farmer sells, he receives $2.13. The farmer sells 108 bushels. How much does the farmer receive?

## Part 8

**Answer the questions.**

a. A rectangular room has an area of 126 square yards. The room is 9 yards wide. How long is the room?

b. A room is 18 feet wide and 21 feet long. What is the perimeter of the room?

c. Each sponge cost $.67. Joan bought 15 sponges. How much did Joan spend?

## Part 9

**Write the fraction for each item.**

a. The denominator is 14. The numerator is 8.

b. The numbers are 12 and 36. The fraction is more than 1.

c. The picture shows 14 parts per unit. 32 parts are shaded.

d. The numbers are 15 and 18. The fraction is less than 1.

# Test 10

## Part 4

| *Price list* | |
|---|---|
| 3 pounds of apples | $1.32 |
| 5 pounds of flour | $ .89 |
| 2 pounds of raisins | $3.27 |
| 6 cans of grape juice | $2 |

a. Julie buys 30 pounds of flour. How much does she spend?

b. Steve spends $12 on grape juice. How many cans does he buy?

c. How much do 34 pounds of raisins cost?

## Part 5

a.

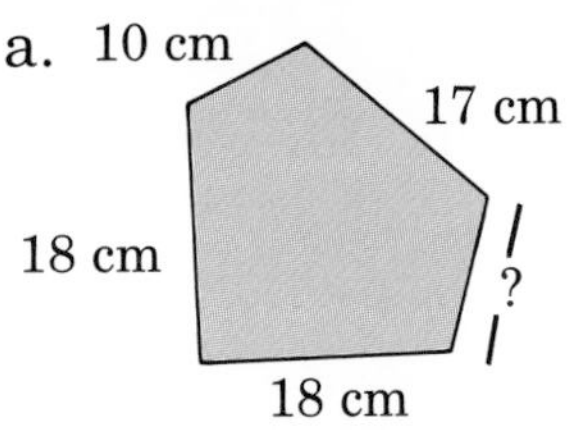


The perimeter is 74 centimeters. What's the length of the missing side?

b.

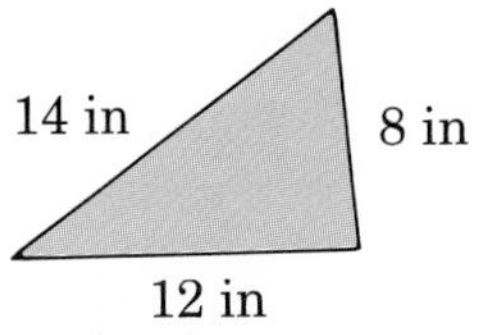


What's the perimeter?

## Part 6

a. For this set, if you took 7 trials you would expect 4 winners. How many trials would you probably have to take to get 56 winners?

b. For this set, if you took 5 trials you would expect 2 winners. How many winners would you expect if you took 85 trials?

## Part 7

a. $\frac{7}{5} \times \blacksquare = \frac{\phantom{0}}{100} = \ .$

b. $\frac{3}{10} \times \blacksquare = \frac{\phantom{0}}{100} = \ .$

c. $\frac{9}{50} \times \blacksquare = \frac{\phantom{0}}{100} = \ .$

## Part 8

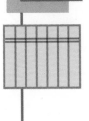

a. A rectangle is 5 inches wide. The area of the rectangle is 185 square inches. How long is the rectangle?

b. A rectangle is 12 inches wide. The rectangle is 24 inches long. What is the area of the rectangle?

# Lesson 100

## Part 1

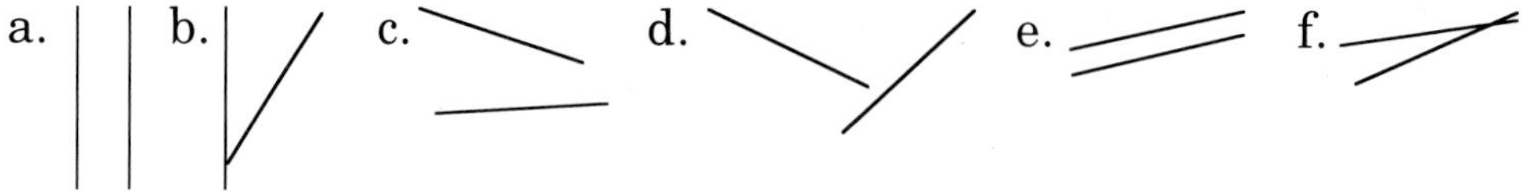


## Part 2

a. $4\frac{1}{2}$ b. $1\frac{5}{8}$ c. $7\frac{4}{7}$ d. $3\frac{2}{50}$

## Part 3

- You can use functions to figure out the "rule" for numbers in a series.
- Here are numbers that follow a rule: **9 8 7 6 5** __ __ __.
- To find the rule, you just call one of the numbers $x$ and figure out what you do to get to the next number.
- The rule for this series is $x - 1$. $x - 1$ tells you how to go from any number in this series to the next number.

# Lesson 101

## Part 1

- It's not easy to see which of these fractions is more: $\frac{5}{4}$ ? $\frac{15}{11}$
- Here's one way to compare them. Start with the fraction with the smaller numbers. $\frac{5}{4} =$
- Write a ratio equation problem with **one** of the numbers from the other fraction.
- Here are the two problems you could try to work: You can work one of these problems.

$$\frac{5}{4} = \frac{15}{\square}$$

$$\frac{5}{4} = \frac{\square}{11}$$

- You can work this problem:

$$\frac{5}{4} \times \frac{3}{3} = \frac{15}{12}$$

$\frac{15}{12}$ **is less than** $\frac{15}{11}$

So: $\frac{5}{4}$ **is less than** $\frac{15}{11}$

a. $\frac{12}{5}$ ? $\frac{3}{2}$

b. $\frac{2}{3}$ ? $\frac{7}{9}$

c. $\frac{5}{3}$ ? $\frac{9}{6}$

## Independent Work

### Part 2

**For each pair of line segments that are parallel, write parallel. For each pair of line segments that intersect, write intersect. For some pairs you won't write anything because they don't intersect in the picture.**

a. 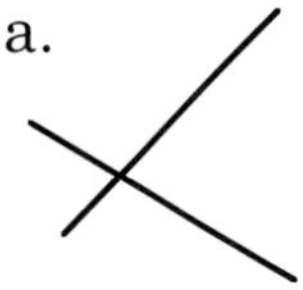

b. 

c. 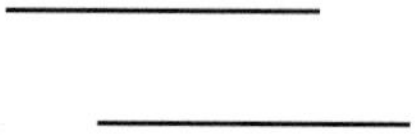

d. 

e. 

f. 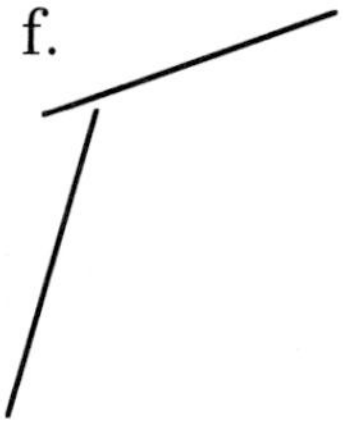

### Part 3

**Copy each problem and work it.**

a. $4\overline{)7036}$

b. $\begin{array}{r} 7034 \\ \underline{\times \quad 4} \end{array}$

c. $\begin{array}{r} 618 \\ \underline{\times \ 35} \end{array}$

d. $3\overline{)2289}$

e. $\begin{array}{r} 6005 \\ \underline{-\ 4363} \end{array}$

### Part J

c. $\frac{\text{winners}}{\text{trials}} \quad \frac{5}{7} \times \frac{20}{20} = \frac{100}{\boxed{140}}$

d. $\frac{\text{winners}}{\text{trials}} \quad \frac{2}{9} \times \frac{50}{50} = \frac{100}{\boxed{450}}$

a. $\frac{\text{winners}}{\text{trials}} \quad \frac{4}{9} \times \frac{2}{2} = \frac{\boxed{8}}{18}$

b. $\frac{\text{winners}}{\text{trials}} \quad \frac{1}{3} \times \frac{150}{150} = \frac{\boxed{150}}{450}$

# Lesson 102

## Part 1

a. $\frac{16}{13}$ ? $\frac{4}{3}$

b. $\frac{2}{3}$ ? $\frac{17}{27}$

c. $\frac{12}{28}$ ? $\frac{2}{5}$

## Part 2

a. For this set, if you took 5 trials, you'd expect to get 4 winners. About how many winners would you expect to get if you took 620 trials?

b.  For this set, the winners are ✘'s. To get 20 winners, about how many trials would you have to take?

c. For this set, if you took 8 trials, you'd expect to get 7 winners. To get 490 winners, about how many trials would you have to take?

## Part 3

**These ratios show how much flour costs at different stores.**

| | Store 1 | Store 2 | Store 3 |
|---|---|---|---|
| $\frac{\text{pounds of flour}}{\$}$ | $\frac{3}{2}$ | $\frac{3}{4}$ | $\frac{3}{1}$ |

**These fractions show how much fuel different tractors use.**

| | Tractor 1 | Tractor 2 | Tractor 3 |
|---|---|---|---|
| $\frac{\text{gallons of fuel}}{\text{miles}}$ | $\frac{2}{15}$ | $\frac{2}{11}$ | $\frac{2}{13}$ |

a. Which store gives the best deal for buying flour?

b. Which store gives the worst deal for buying flour?

c. Which tractor gets the best gas mileage?

d. Which tractor gets the worst mileage?

**Part J**

a. $\frac{\text{winners}}{\text{trials}} \quad \frac{4}{5} \times \frac{124}{124} = \frac{\boxed{496}}{620}$

b. $\frac{\text{winners}}{\text{trials}} \quad \frac{1}{8} \times \frac{20}{20} = \frac{20}{\boxed{160}}$

c. $\frac{\text{winners}}{\text{trials}} \quad \frac{7}{8} \times \frac{70}{70} = \frac{490}{\boxed{560}}$

I got a winner.
I got a winner.
I think I got a loser.

# Lesson 103

## Part 1

a. $\frac{20}{27} \; ? \; \frac{4}{7}$ b. $\frac{5}{3} \; ? \; \frac{26}{18}$ c. $\frac{20}{152} \; ? \; \frac{1}{8}$

## Part 2

These fractions tell about the cost of tablets at different stores.

| | Store 1 | Store 2 | Store 3 |
|---|---|---|---|
| $\frac{\text{tablets}}{\$}$ | $\frac{7}{3}$ | $\frac{7}{5}$ | $\frac{7}{2}$ |

These ratios tell about how much plant food different farmers used.

| | Farmer 1 | Farmer 2 | Farmer 3 |
|---|---|---|---|
| $\frac{\text{gallons of plant food}}{\text{acres}}$ | $\frac{4}{3}$ | $\frac{1}{3}$ | $\frac{5}{3}$ |

***Questions***

a. Which store gives the best deal for buying tablets?

b. Which store gives the worst deal for buying tablets?

c. What is the fraction for the worst deal for buying tablets?

d. Which farmer has the best ratio for the number of gallons and the number of acres treated?

e. Which is the worst fraction for the number of gallons and the number of acres treated?

# Independent Work

**Part 3** **For each row, write the multiplication equation, the division problem and the answer, the fraction equation and the division equation.**

| | □ x __ = □ | ⟌ | $\frac{□}{□}$ = □ | □ ÷ □ = |
|---|---|---|---|---|
| a. | 7 x __ = 49 | | | |
| b. | | | | 36 ÷ 4 = |
| c. | | | | 87 ÷ 3 = |
| d. | | | $\frac{92}{2}$ = | |

**Part 4** **Copy the table. Figure out the function and complete the table.**

| | | Function | |
|---|---|---|---|
| | $x$ | | $y$ |
| a. | 3 | | 9 |
| b. | 5 | | 11 |
| c. | 0 | | |
| d. | 2 | | |

**Part 5** **For each row, write the decimal number, the fraction and the mixed number.**

| | □.□ | $\frac{□}{□}$ | □ + $\frac{□}{□}$ |
|---|---|---|---|
| a. | 4.07 | | |
| b. | | | $20 + \frac{6}{100}$ |
| c. | | $\frac{372}{100}$ | |
| d. | 18.60 | | |

**Part 6** **Write the subtraction fact or division fact for each set of numbers.**

a. 2, 8, 4 b. 16, 4, 12 c. 8, 4, 32 d. 1, 1, 1

# Lesson 104

**Part 1**

***Sample problem:*** At Rita's store, you can buy 5 pounds of flour for \$2. At Andy's store, you can buy 13 pounds of flour for \$6. At which store does flour cost less?

a. At M-Mart, you can buy 7 pens for \$2. At Z-Mart, you can buy 18 pens for \$4. At which store do pens cost less?

b. At M-Mart, 9 pounds of nuts cost \$23. At Z-Mart, 3 pounds of nuts cost \$8. At which store do nuts cost less?

# Lesson 105

## Part 1

a. At M-Mart, 9 pounds of beans cost $4. At Z-Mart, 27 pounds of beans cost $13. At which store do the beans cost less?

b. Mow More mowers run for 12 hours on 7 pints of gas. Cutter mowers run for 3 hours on 2 pints of gas. Which mower uses less gas?

## Independent Work

## Part 2

**Write the fraction for each item.**

a. The numerator is 17. The denominator is 42.

b.  What are your chances of drawing an ✘ on the first trial?

c. The fraction is more than 1. The numbers are 14 and 9.

d. The picture shows 20 parts per unit and 100 parts are shaded.

e. 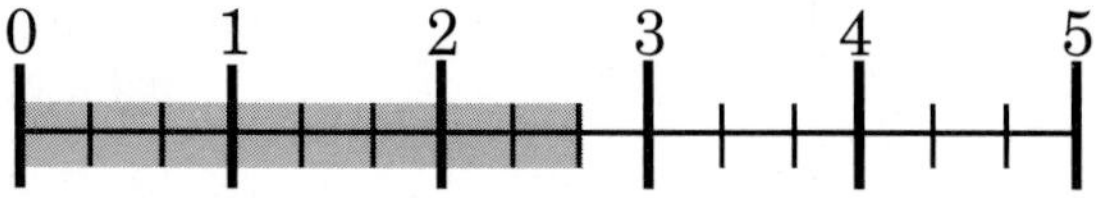

## Part 3

**Some of these problems ask about the area or tell about the area. Some of these problems ask about the perimeter or tell about the perimeter. Read each problem carefully and answer the question.**

a.

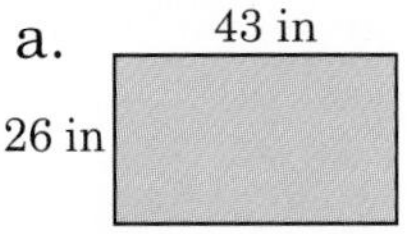


What is the perimeter of figure a?

b.

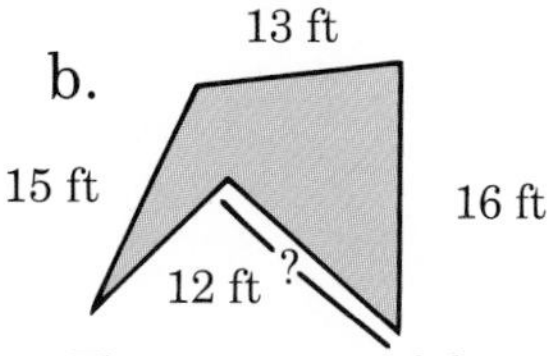


The perimeter of the figure is 70 feet. What is the length of the missing side?

c.

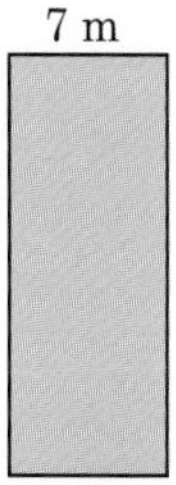


The area of the rectangle is 133 square meters. How high is the rectangle?

d. What is the area of rectangle a?

## Part 4

**Write parallel if the line segments are parallel. Write intersect if the line segments intersect.**

a.

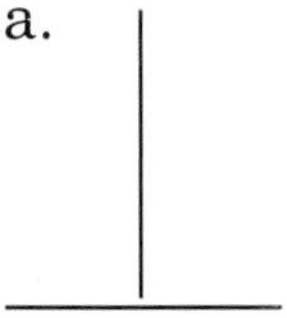

b.

c.

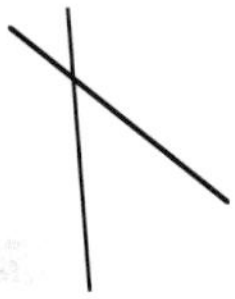

d.

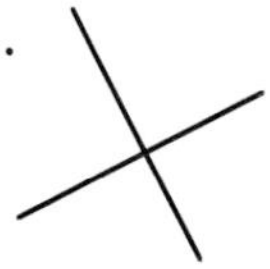

e.

f.

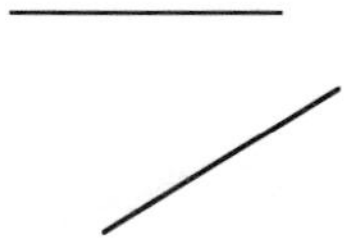

## Part 5

**Copy each problem and work it.**

a. $7 - \frac{14}{3} =$

b. $5 \times \frac{2}{8} =$

c. $\frac{13}{2} + 4 =$

d. $\frac{5}{4} \times 11 =$

## Part J

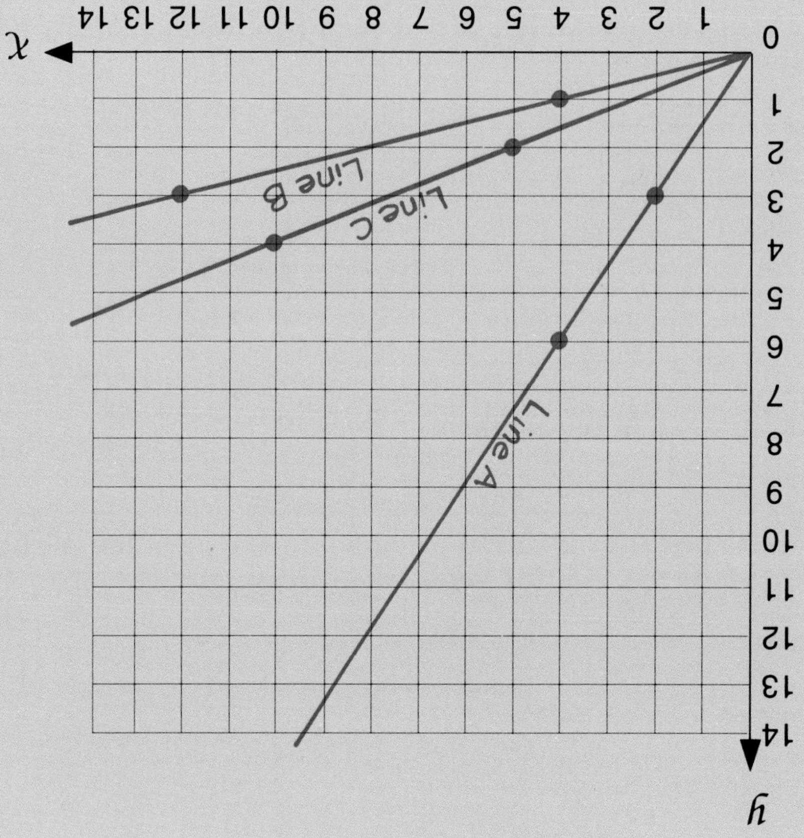

# Lesson 106

## Part 1

- You've put fractions on a coordinate grid. The denominator of the fraction is the $x$ value. The numerator of the fraction is the $y$ value.
- You can draw lines on the coordinate grid to show equivalent fractions. That means you can draw lines on the coordinate grid to show answers to **ratio problems.**
- Here's a ratio problem: The ratio of hands to people is 2 to 1. If there are 5 people, how many hands are there?
- Here's the ratio equation with names and numbers the problem gives: $\frac{\textbf{hands}}{\textbf{people}}\ \frac{2}{1} = \frac{\blacksquare}{5}$
- For this problem, people is the name for the $x$ axis. Hands is the name for the $y$ axis.
- Here is the point for the first fraction and the line:

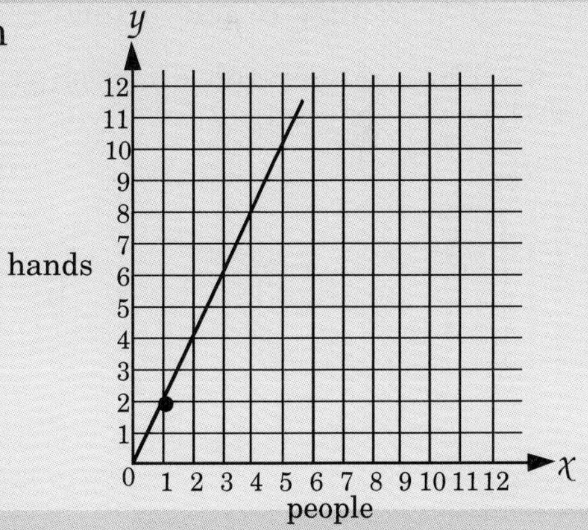


- The line goes through the number of hands for 5 people. Go from 5 on the $x$ axis to the line. Then go over to the $y$ axis.
- If there are 5 people, there are 10 hands.

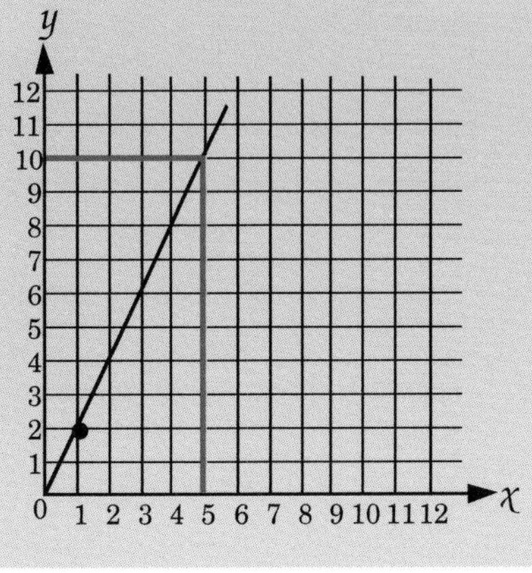

a. A recipe calls for 3 cups of onions for every 2 tablespoons of garlic. A cook follows the same ratio, but uses 9 cups of onions. How many tablespoons of garlic will the cook need?

b. A set has winners and losers. For the set, if you took 4 trials, you would expect to get 1 winner. About how many trials would you have to take to get 3 winners?

## Part 2

- Some division problems have a remainder after the last digit of the answer.
- Here's a problem: $5\overline{)39}$
- 35 goes below 39.

$$\begin{array}{r} 7 \\ 5\overline{)39} \\ \underline{35} \end{array}$$

- There's a remainder of 4. That remainder is written after the 7 in the answer.
- The answer tells you that 39 = 5 x 7 + a remainder of 4.

$$\begin{array}{r} 7 + R4 \\ 5\overline{)39} \\ \underline{35} \end{array}$$

a. $3\overline{)13}$   b. $9\overline{)15}$   c. $9\overline{)30}$   d. $7\overline{)19}$

## Part 3

***Sample problem:*** $\frac{3}{4}$ ? .70

a. $\frac{7}{5}$ ? 1.46   b. $\frac{9}{21}$ ? .36

## Independent Work

### Part 4

**Copy each item and write the mixed number it equals or the whole number it equals. For the mixed numbers, do not write a plus sign between the whole number and the fraction.**

a. $\frac{18}{5} =$ b. $\frac{36}{4} =$ c. $\frac{75}{4} =$ d. $\frac{75}{3} =$

### Part 5

**Copy each pair of fractions. Below, write the ratio equation that starts with the smaller fraction and has one of the numbers from the other fraction. Then circle the fraction that is larger.**

a. $\frac{17}{36}$ ? $\frac{4}{9}$ b. $\frac{42}{11}$ ? $\frac{7}{2}$ c. $\frac{5}{6}$ ? $\frac{19}{24}$

### Part 6

**Complete each equation.**

a. $\frac{5}{3} \times \blacksquare = \frac{15}{24}$ b. $\frac{\blacksquare}{7} \times \frac{9}{4} = \frac{45}{28}$ c. $\frac{11}{\blacksquare} \times \frac{5}{7} = \frac{55}{14}$

### Part J

a. $\frac{1}{3} \times \frac{32}{32} = \frac{32}{96}$ $\frac{32}{100}$

b. $\frac{5}{6} \times \frac{17}{17} = \frac{85}{102}$ $\frac{85}{100}$

c. $\frac{6}{5} \times \frac{20}{20} = \frac{120}{100}$ $\frac{109}{100}$

# Lesson 107

## Part 1

a. $7\overline{)25}$ b. $5\overline{)373}$ c. $7\overline{)109}$ d. $7\overline{)149}$

## Part 2

a. $\frac{1}{3}$ ? .32 b. $\frac{5}{6}$ ? .85 c. 1.09 ? $\frac{6}{5}$

## Part 3

- You can use lines on the coordinate grid to answer questions.
- Here's the line for the number of people and the number of hands:
- The ratio of hands to people is 2 to 1.
- The line goes through the point for zero.
- Here are some question u can answer.

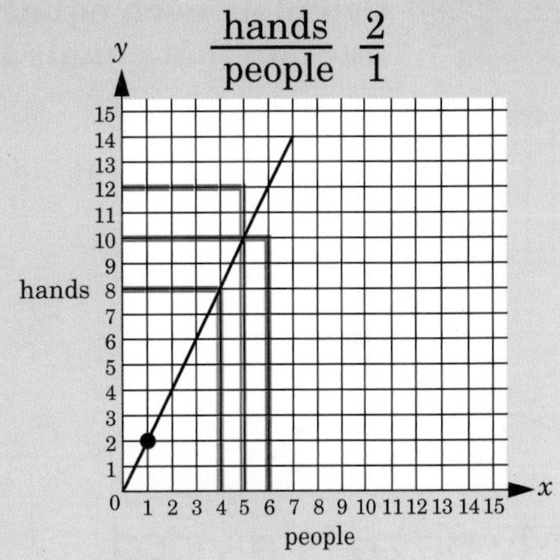


- 1. How many hands do 6 people have?

  Go to 6 on the $x$ axis. Follow the arrows at 6 up to the line and over to the $y$ axis. The number on the $y$ axis shows the number of hands. For 6 people there are 12 hands.

- 2. How many hands do 4 people have?

  Go to 4 on the $x$ axis. Follow the arrows at 4 up to the line and over to the $y$ axis. The number on the $y$ axis shows the number of hands. For 4 people there are 8 hands.

- 3. If there are 10 hands, how many people are there?

  Go to 10 on the $y$ axis. Follow the arrows, at 10 over to the line and down to the $x$ axis. The number on the $x$ axis shows the number of people. For 10 hands there are 5 people.

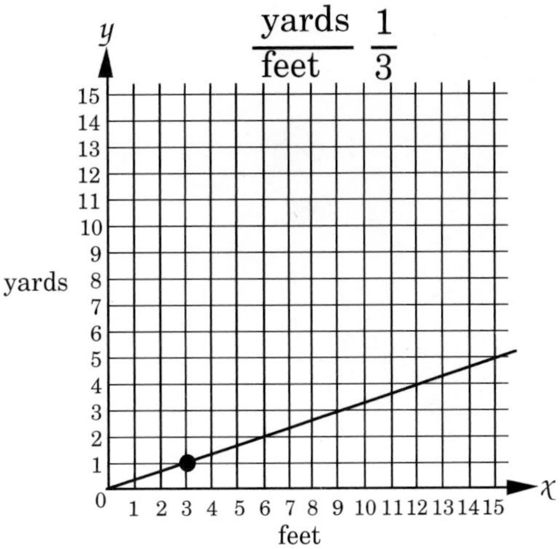


a. How many yards is 9 feet?

b. How many yards is 6 feet?

c. How many feet is 4 yards?

## Independent Work

**Part 4** **Complete each equation to show the missing number and the fraction that equals 1.**

a. $\frac{1}{5} = \frac{\blacksquare}{85}$  b. $\frac{4}{3} = \frac{428}{\blacksquare}$  c. $\frac{\blacksquare}{5} = \frac{18}{30}$  d. $\frac{9}{\blacksquare} = \frac{27}{12}$

# Lesson 108

## Part 1

a. $9\overline{)129}$ b. $4\overline{)727}$ c. $3\overline{)743}$ d. $5\overline{)1204}$

## Part 2

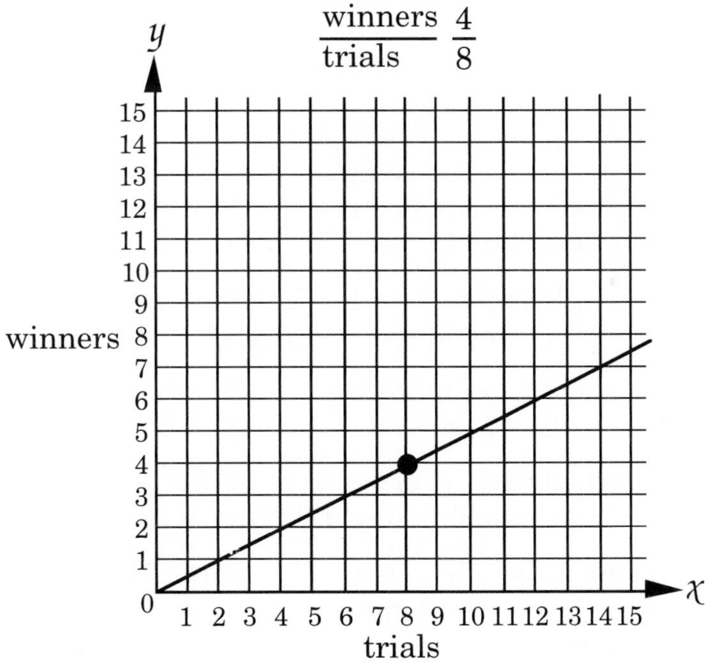


a. If you took 10 trials, how many winners would you expect?

b. If you took 4 trials, how many winners would you expect?

c. If you took trials until you got 6 winners, about how many trials would you take?

d. If you took trials until you got 2 winners, about how many trials would you take?

e. If you took 6 trials, how many winners would you expect?

f. If you took trials until you got 4 winners, about how many trials would you take?

## Independent Work

**Part 3** **For items a and c, use a ratio equation to compare the fractions. Then answer all the questions.**

a. Is $\frac{7}{3}$ more or less than $\frac{63}{29}$?

b. Is $\frac{63}{29}$ more or less than $\frac{63}{26}$?

c. Is $\frac{7}{3}$ more or less than $\frac{63}{26}$?

d. Which fraction is closer to $\frac{63}{26}$, $\frac{7}{3}$ or $\frac{63}{29}$? (***Hint:*** Use the information from items a, b and c.)

**Part 4** **For each problem, figure out the answer and box it. For some problems, make a number family. For other problems, write a ratio equation. For the rest of the problems, draw a diagram.**

a. A rectangle is 9 feet wide. The rectangle is 36 feet high. What is the area of the rectangle?

b. Jim earned $164 more than Buffy. Jim earned $981. How much did Buffy earn?

c. Jim earned $164 every 3 days he worked. Jim worked 33 days. How much money did Jim earn?

d. Jim had $164. He earned $87 more. His mother gave him another $75 to go shopping. Jim spent some money at the store. Jim ended up with $132. How much money did Jim spend?

e. There were 42 birds in the marsh. 8 were geese and the rest were ducks. What fraction of the birds in the marsh were ducks?

f. There were geese and ducks in the marsh. There were 8 geese for every 5 ducks. There were 80 ducks. How many geese were in the marsh?

g. The perimeter of a 5-sided figure is 112 centimeters. Here are the lengths of 4 sides: 24 centimeters, 20 centimeters, 29 centimeters and 25 centimeters. What's the length of the missing side?

## Part 5

**Write the mixed number or the whole number for each fraction.**

a. $\frac{77}{3} =$ b. $\frac{77}{7} =$ c. $\frac{77}{4} =$ d. $\frac{77}{5} =$

## Part 6

**Complete each equation. Remember to show the fraction that equals 1 for the equivalent fractions.**

a. $\frac{2}{7} \times \frac{\blacksquare}{3} = \frac{24}{\blacksquare}$ b. $\frac{14}{5} \times \frac{3}{6} = \blacksquare$ c. $\frac{3}{4} = \frac{21}{\blacksquare}$

d. $\frac{9}{2} = \frac{\blacksquare}{18}$ e. $\frac{3}{\blacksquare} = \frac{27}{63}$ f. $\frac{\blacksquare}{5} = \frac{28}{20}$

## Part 7

**Complete each equation. Write the missing value as a mixed number.**

a. $\frac{\blacksquare}{4} \times \frac{5}{5} = \frac{17}{20}$ b. $\frac{9}{\blacksquare} \times \frac{7}{7} = \frac{63}{39}$

Don't you know the rule about multiplying by 1?
Sure. I just can't remember which one to multiply by.

# Lesson 109

## Part 1

a. $6\overline{)38}$ b. $9\overline{)545}$ c. $6\overline{)570}$ d. $7\overline{)222}$

## Independent Work

## Part 2

**For each problem, write a ratio equation or number family or draw a diagram. Then figure out the answer.**

a. A 6-sided figure has a perimeter of 97 feet. The sides that are shown have lengths of 8 feet, 12 feet, 19 feet, 20 feet and 17 feet. What is the length of the side that is not shown?

b. John bought 5 pounds of apples for $2. How many pounds of apples could John buy for $70?

c. John had 57 pounds of apples. John sold 12 pounds of apples. John cooked pies with 6 pounds of apples. John ate another 11 pounds of apples. How many pounds of apples did John end up with?

d. At a store, there were 3 apples per pound. If there were 60 pounds of apples, how many apples were there?

e. A conference room was 18 meters long and 15 meters wide. What was the area of the room?

f. A crate contained green apples and red apples. There were 86 apples in the crate. 27 of the apples were red. How many apples were green?

g. In problem f, what fraction of the apples were red?

h. In problem f, what is the fraction for your chances of pulling an apple from the crate?

**Part 3** **For each row, write the multiplication equation, the division problem and the answer, the fraction equation and the division equation.**

| | □ x __ = □ | $\overline{)\square}$ | $\frac{\square}{\square} = \square$ | □ ÷ □ = |
|---|---|---|---|---|
| a. | | | $\frac{75}{5} =$ | |
| b. | | | | 12 ÷ 3 = |
| c. | | $1\overline{)4}$ | | |
| d. | 9 x __ = 144 | | | |

**Part 4** **On the coordinate grid, write the names for the ratio on the** $x$ **axis and the** $y$ **axis. Plot the point for the fraction and draw the line for the function. Use the line to answer each question.**

$$\frac{\text{winners}}{\text{trials}} \quad \frac{2}{3}$$

a. How many winners would you expect if you took 12 trials?

b. About how many trials would you take to get 12 winners?

c. About how many trials would you take to get 6 winners?

d. How many winners would you expect if you took 18 trials?

## Part J

a. 6⟌38 = 6 + R2 (36)

b. 9⟌545 = 60 + R5 (0)

c. 6⟌570 = 95 (54, 30)

d. 7⟌222 = 31 + R5 (21, 7)

## Part K

a. 7⟌529 = 75 + R4 (49, 35)

b. 6⟌842 = 140 + R2 (6, 00)

c. 7⟌45 = 6 + R3 (42)

d. 4⟌2006 = 501 + R2

If each car holds 4 people, we'll have 3 full cars.
And we're the remainder.

# Lesson 110

## Part 1

a. $7\overline{)529}$ b. $6\overline{)842}$ c. $7\overline{)45}$ d. $4\overline{)2006}$

## Independent Work

## Part 2

**Answer the question for each item.**

a.

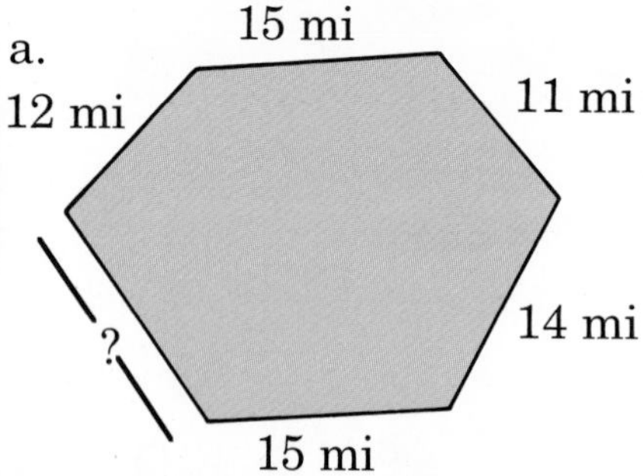


The perimeter of figure a is 82 miles. What is the length of the missing side?

b.

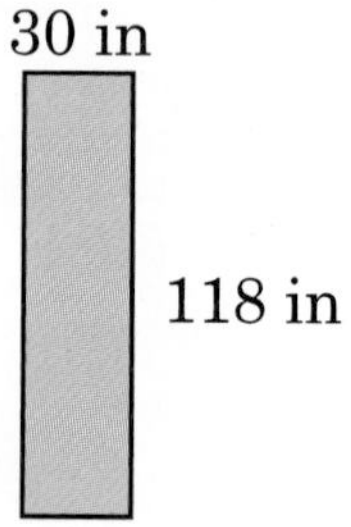


What is the area of figure b?

c. What is the perimeter of figure b?

d.

What is the height of figure d?

e. What is the perimeter of figure d?

**Part 3** Complete each equation.

a. $\frac{7}{3} \times \frac{8}{■} = \frac{■}{21}$

b. $\frac{16}{5} - \frac{16}{5} = ■$

c. $\frac{8}{9} \times 4 = ■$

d. $\frac{34}{5} + 7 = ■$

e. $\frac{6}{5} \times ■ = \frac{30}{25}$

f. $3 \times ■ = 87$

g. $48 \div 4 = ■$

# Lesson 111

## Independent Work

**Part 1** **For each item, write both values as a fraction. Then make a ratio equation that starts with the fraction having the smaller numbers. Complete the equation with one of the numbers from the other fraction. Circle the fraction that is larger.**

a. $.40 \; ? \; \frac{5}{12}$

b. $\frac{3}{2} \; ? \; \frac{79}{52}$

c. $1\frac{7}{12} \; ? \; \frac{7}{4}$

**Part 2** **Work each item.**

a.

What is the width of figure a?

b.

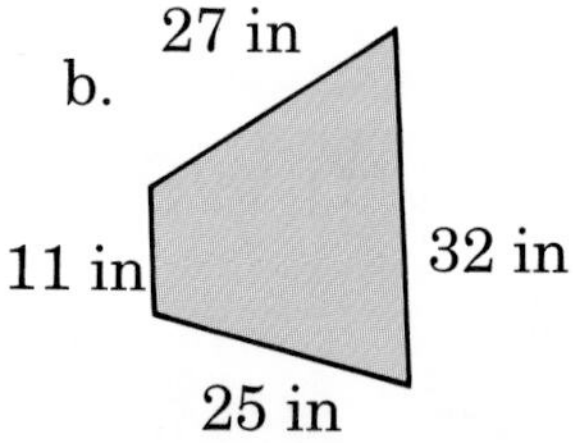


What is the perimeter of figure b?

c. Use the answer to item a and figure out the perimeter of figure a.

d. Which figure has only one pair of parallel sides?

# Lesson 112

## Independent Work

**Part 1** **For each problem, write the names and ratios you'll compare. Complete the equation for the fraction that has the smaller numbers. Then answer the question.**

a. Car A travels 279 miles on 9 gallons of fuel. Car B travels 29 per gallon. Which car has the best mileage?

b. For bag A, if you took 13 trials you would expect 6 winners. For bag B, your chances for drawing a winner is .48. Which bag gives you the best chance of drawing a winner? (***Hint:*** Write the fraction for bag B. Then compare the fractions.)

**Part 2** **Answer each question.**

a. The perimeter of a figure is 106 inches. The sides that are shown have lengths of 22, 18, 20 and 23 inches. What is the length of the missing side?

b. There are cars and trucks in a lot. There are 72 more trucks than cars. There are 147 trucks. How many cars are there?

c. There are cars and trucks in a lot. There are 72 trucks in the lot. There are 147 vehicles in the lot. What fraction of the vehicles are cars?

d. Jim's room is 12 feet wide and 18 feet long. What is the area of Jim's room?

e. New carpet costs $1.34 per square foot. Use the answer to problem d and find out how much it would cost to put new carpet in Jim's room.

f. A man had $174 in the bank. He received a pay check for $425. He got a bonus of $50. He spent some money on a bicycle. He ended up with $562. How much money did the bicycle cost?

# Lesson 113

## Independent Work

**Part 1** **Answer each question.**

a. The sides of a figure are 17, 12, 14, 9, 6 and 15 yards long. What is the perimeter of the figure?

b. 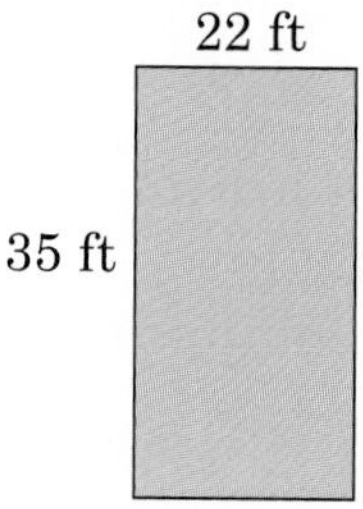

What is the area of figure b?

c. A hamburger cost $2.35. John bought 6 hamburgers. How much did John spend?

d. A tank held some water. 54 gallons were poured out of the tank. 77 gallons leaked out of the tank. 2 gallons evaporated from the tank. 133 gallons were left in the tank. How many gallons did the tank have to start with?

e. In store A, 6 pounds of rice cost $2.80. How much would 24 pounds of rice cost?

f. At store B, 4 pounds of rice cost $1.87. In problem e you figured out how much 24 pounds of rice cost at store A. Use that information to figure out which store has a better deal.

g. Store R sells chicken for $.38 less a pound than store Q. Store Q sells chicken for $1.94 a pound. How much does store R sell chicken for?

h. In Cornflat there are two stores– Joe's Market and Tim's Shop. The two stores sold 803 cans of soup. Tim's Shop sold 198 cans of soup. How many cans of soup did Joe's Market sell?

# Lesson 114

## Independent Work

**Part 1** **Answer each question.**

a. A man walks 12 miles in 3 hours. How many miles per hour does the man walk?

b. A hardware store gives away 5 feet of chain for every 2 rolls of insulation a customer purchased. How many feet of chain would the store give to a customer who purchased 32 rolls of insulation?

c. A gardening shop gives away 8 feet of chain for every 3 rolls of insulation purchased. Use the information about 32 rolls of insulation from problem b and figure out if the hardware store or the gardening shop has a better deal.

d. A cow weighed 185 pounds less than a bull. The cow weighed 1174 pounds. How much did the bull weigh?

**Part 2** **Copy the table. Write the $x$ and $y$ values for each point and figure out the function.**

| | $x$ | Function | $y$ |
|---|---|---|---|
| A | | | |
| B | | | |
| C | | | |

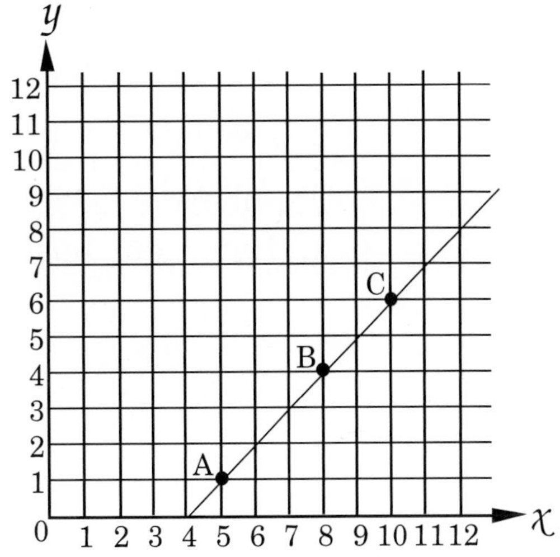

**Part 3** **Use the line to answer the questions.**

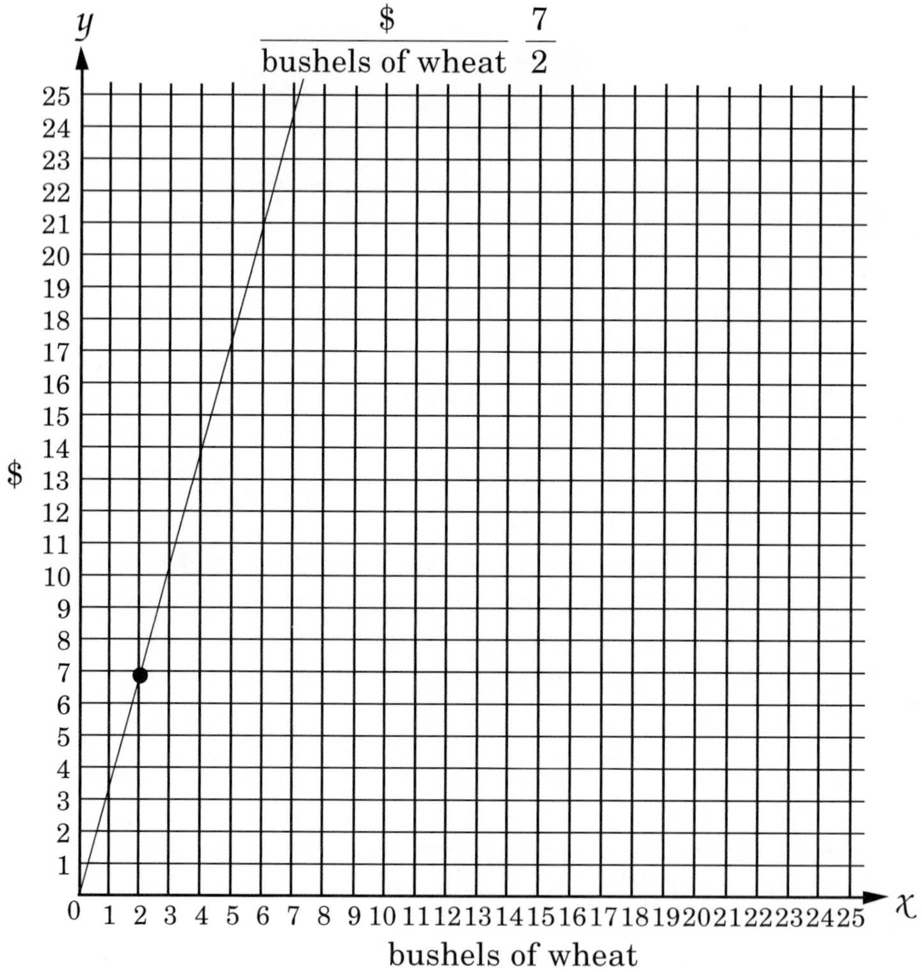


a. How many bushels of wheat cost $14?

b. About how much money does 3 bushels of wheat cost?

c. About how many bushels of wheat can you buy for $10?

d. How many bushels of wheat cost $21?

# Lesson 115

## Part 1

Two farmers had an argument about whether it was smarter to fence small fields or big fields.

Farmer Jones said, "It's smarter to fence small fields. For small fields, you use less fence compared to the area inside the field."

Farmer Black said, "No way. If you want to use less fencing and fence more area, you should fence a very large field, not a small one."

They argued over this issue for a long time. Finally, they decided to make a test. Both of them agreed to fence a square field and to figure out the amount of fencing and the area of the land that was fenced. They decided to measure everything in rods. (A rod is a unit of length that is $16\frac{1}{2}$ feet long.)

Farmer Jones fenced a very small field. It was a square field that was 3 rods on each side.

Farmer Black fenced a larger field. It was also a square, but it was 9 rods on each side.

When they were all done, they had to figure out who had the best ratio for the amount of fencing to the area inside the fence.

- Here's the ratio equations and the names:

$$\frac{\textbf{Jones' fencing}}{\textbf{Jones' area fenced}} \quad \boxed{\frac{\quad}{\quad}}$$

$$\frac{\textbf{Black's fencing}}{\textbf{Black's area fenced}} \quad \boxed{\frac{\quad}{\quad}}$$

## Independent Work

**Part 2** **Use the line to work the items.**

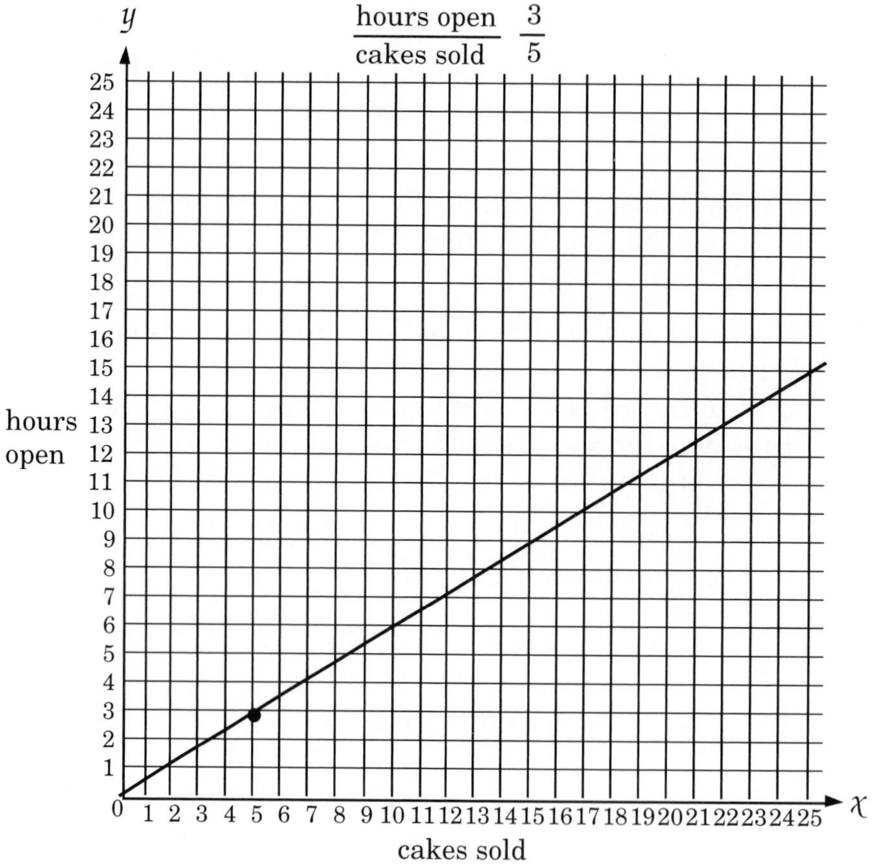


a. This graph shows the sales at a bakery. Write the names and the ratio for the bakery.

b. The bakery was open 9 hours on Saturday. How many cakes did the bakery sell?

c. If the bakery sold 20 cakes, how many hours was the bakery open?

d. Write a ratio equation to figure out how many cakes the bakery would sell in 75 hours. Box the answer.

e. A grocery store sells cakes for 24 hours. In that time, the grocery store sells 32 cakes. Does the bakery or the grocery store sell more cakes per hour?

## Part 3 Copy the table. Write the $x$ and $y$ values for each point and figure out the function.

| | $x$ | Function | $y$ |
|---|---|---|---|
| A | | | |
| B | | | |
| C | | | |

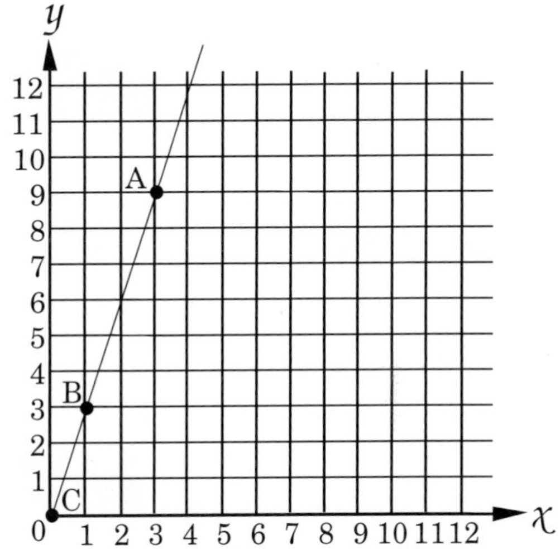


## Part 4 Answer the question.

a. One number is 37 more than another number. The larger number is 37. What's the smaller number?

b. Two numbers added together equal 194. One of the numbers is 117. What's the other number?

## Part 5 Answer the questions. Remember the equation:

$$\textbf{area of triangle} = \frac{\textbf{area of rectangle}}{\textbf{2}}$$

a.

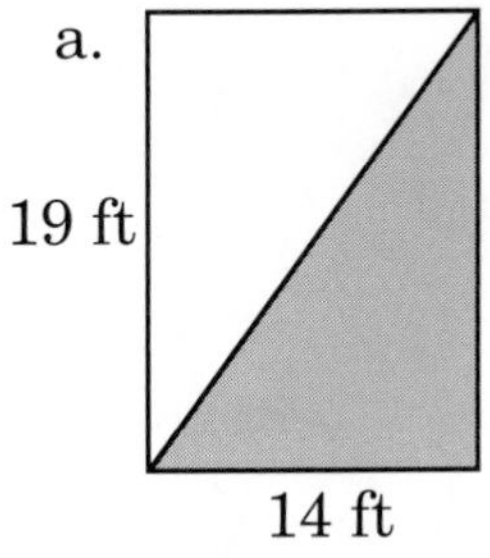


What is the area of rectangle a?

b. What is the area of the shaded triangle?

c. What is the area of the unshaded triangle?

d. What is the perimeter of the rectangle?

# Lesson 116

## Independent Work

**Part 1** **Complete each equation.**

a. $17 = \frac{\phantom{0}}{1} = \frac{\phantom{0}}{4} = \frac{\phantom{0}}{25}$

b. $\frac{4}{9} \times \frac{13}{\blacksquare} = \frac{\blacksquare}{36}$

c. $\frac{15}{2} - 4 = \blacksquare$

d. $81 \div 9 = \blacksquare$

e. $\frac{2}{5} \times 4 = \blacksquare$

f. $7 \times \blacksquare = 56$

g. $\frac{7}{3} + \blacksquare = 4$

h. $200 \div 5 = \blacksquare$

I think we're close to the end.

# Lesson 118

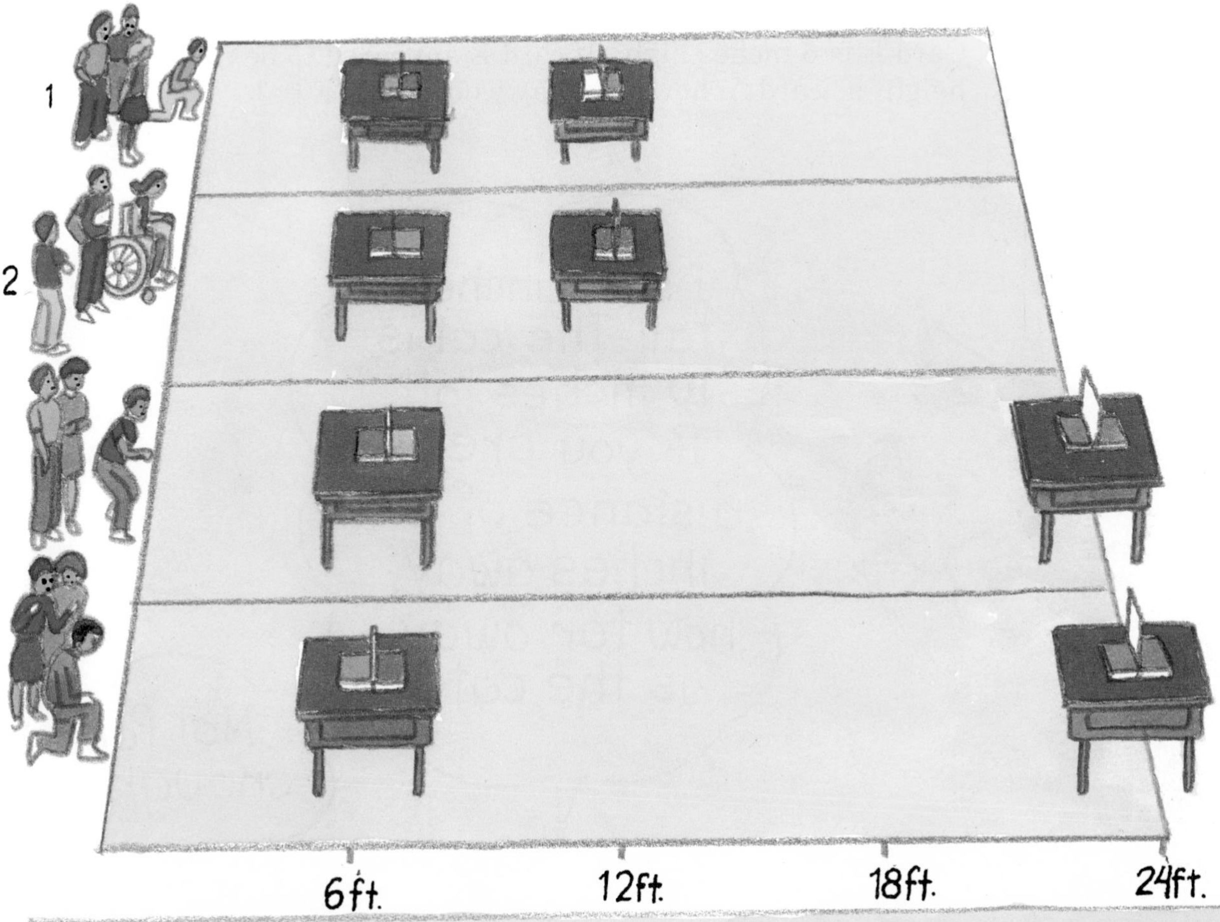


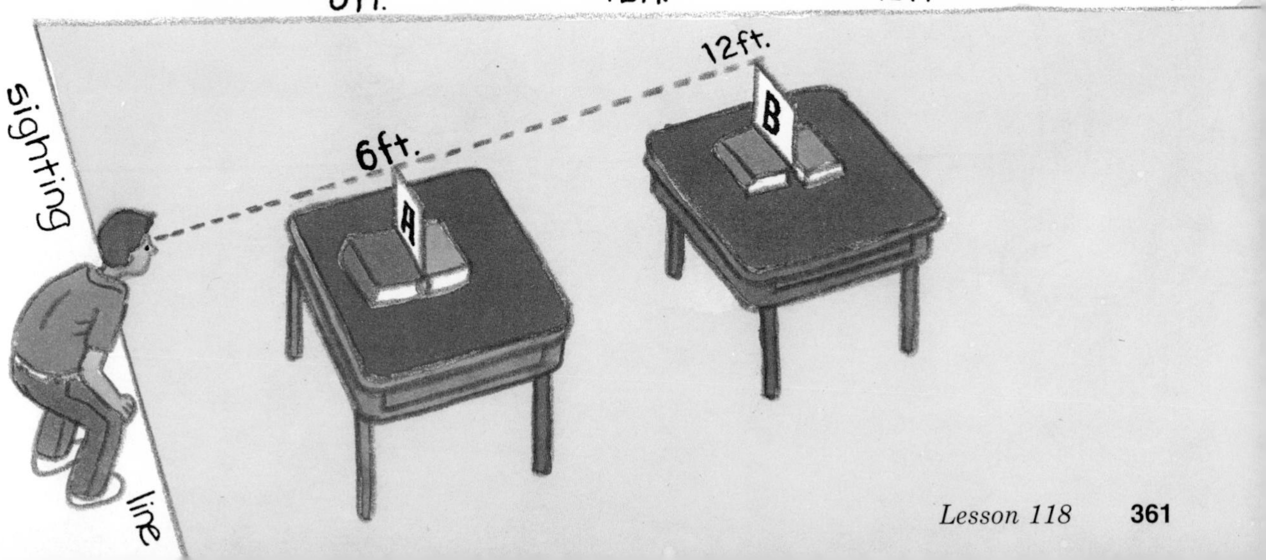

# Lesson 119

**Part 1**

a. Card D was 15 feet away and appeared to be the same height as card A. What is the height of card D?

b. Card E is 6 inches high. If card E appeared to be the same height as card A, how far away would card E be?

# Lesson 120

**Cake**
**(serves 4)**

Heat oven to 350°F.

Grease and flour a cake pan.

Mix together:
- 2 cups sifted flour
- $\frac{5}{2}$ teaspoons baking powder
- $\frac{1}{8}$ teaspoon salt

In a separate bowl mix together:
- $\frac{1}{4}$ cup shortening
- 1 cup sugar (slowly)
- $\frac{3}{4}$ cup milk
- 1 teaspoon vanilla
- 1 egg

Mix contents of both bowls together.

Pour into a cake pan.

Bake *50 minutes or until done.*

Function

| | | | **x** | |
|---|---|---|---|---|
| | People | 4 | | 12 |
| Ingredients | Flour | 2 | | cups |
| | Baking Powder | $\frac{5}{2}$ | | teaspoons |
| | Salt | $\frac{1}{8}$ | | teaspoon |
| | Shortening | $\frac{1}{4}$ | | cup |
| | Sugar | 1 | | cups |
| | Eggs | 1 | | |
| | Milk | $\frac{3}{4}$ | | cups |
| | Vanilla | 1 | | teaspoons |

## Part J

| | | | Function $x \times 3$ | |
|---|---|---|---|---|
| | People | 4 | $4 \times 3$ | 12 |
| Ingredients | Flour | 2 | $2 \times 3$ | 6 cups |
| | Baking Powder | $\frac{5}{2}$ | $\frac{5}{2} \times \frac{3}{1}$ | $\frac{15}{2}$ teaspoons |
| | Salt | $\frac{1}{8}$ | $\frac{1}{8} \times \frac{3}{1}$ | $\frac{3}{8}$ teaspoon |
| | Shortening | $\frac{1}{4}$ | $\frac{1}{4} \times \frac{3}{1}$ | $\frac{3}{4}$ cup |
| | Sugar | 1 | $1 \times 3$ | 3 cups |
| | Eggs | 1 | $1 \times 3$ | 3 |
| | Milk | $\frac{3}{4}$ | $\frac{3}{4} \times \frac{3}{1}$ | $\frac{9}{4}$ cups |
| | Vanilla | 1 | $1 \times 3$ | 3 teaspoons |